Advances in Nanosheets - Preparation, Properties and Applications

Edited by Karthikeyan Krishnamoorthy

Published in London, United Kingdom

Advances in Nanosheets - Preparation, Properties and Applications
http://dx.doi.org/10.5772/10.5772/intechopen.103990
Edited by Karthikeyan Krishnamoorthy

Contributors
Sita Kandel, Godfrey Gumbs, Oleg L. Berman, Hadja Imane Beloufa, Mostapha Tarfaoui, Khalid Lafdi, Mohamed Daly, Amine Bendarma, Kamlesh Paswan, Somnath Chattopadhyaya, Anil Dube, Mostafa Khosroupour Arabi, Morteza Ghorbanzadeh Ahangari, Karthikeyan Krishnamoorthy, Parthiban Pazhamalai, Rajavarman Swaminathan, Sang-Jae Kim, Barsha Das, Sagnik Das, Soumyabrata Tewary, Sujoy Bose, Sandip Ghosh, Avijit Ghosh, Sasireka Rajendran, Vinoth Rathinam, Vasanth Kumar, Manucroo Kandasamy, Sharmila Selvi Muthuvel, Shanmugasundari Arumugam

First published in London, United Kingdom, 2023 by IntechOpen
IntechOpen is the global imprint of INTECHOPEN LIMITED, registered in England and Wales, registration number: 11086078, 5 Princes Gate Court, London, SW7 2QJ, United Kingdom
Printed in Croatia

British Library Cataloguing-in-Publication Data
A catalogue record for this book is available from the British Library

Additional hard and PDF copies can be obtained from orders@intechopen.com

Advances in Nanosheets - Preparation, Properties and Applications
Edited by Karthikeyan Krishnamoorthy
p. cm.

This title is part of the Nanotechnology and Nanomaterials Book Series, Volume 1
Topic: Nanomaterials and Nanostructures
Series Editor: Jung Huang
Topic Editor: Alberto Jiménez-Suárez
Associate Topic Editor: Gonzalo Seisdedos

Print ISBN 978-1-83769-517-1
Online ISBN 978-1-83769-516-4
eBook (PDF) ISBN 978-1-83769-518-8
ISSN 3029-0538

IntechOpen Book Series

Nanotechnology and Nanomaterials

Volume 1

Aims and Scope of the Series

Humans face growing challenges in environmental sustainability. We need better materials, more effective devices, and higher computing power to conquer these challenges. Researchers have long explored the realm of science at the nanometer scale for developing novel problem-solving technologies. In the past decades, breakthroughs in multiscale computer simulation, novel hierarchical material and device structures, nanometrology, and new functionalities from interfacial and quantum size effects have enhanced our ability to utilize nanotechnology to solve real-world problems. This book series addresses important advancements in nanotechnology and nanomaterials. It showcases how nanotechnology is being continually developed and implemented in a variety of domains of science and technology. The two main topics this series covers are Nanotechnology and Nanodevices, and Nanomaterials and Nanostructures.

Meet the Series Editor

Jung Y. Huang, a university educator and researcher, has been working to unravel the structures and functional properties of materials and cellular events in living cells with varying optical methodologies. He has co-authored hundreds of journal papers and five book chapters. He also holds tens of patents in laser techniques and single-molecule/hyperspectral imaging and has developed architectural photonics based on hierarchically structured materials. Currently, his research focuses on the integration of artificial intelligence methodology with optics to automatically discover meaningful information from optical sensing/imaging data cubes. He has been an editorial board member and reviewer for several scientific journals. As a member of the global scientific community, he sincerely supports and endeavors to promote the spread of scientific knowledge.

Meet the Volume Editor

Dr. Karthikeyan Krishnamoorthy is an Associate Professor in the Department of Physics at the School of Advanced Sciences, Vellore Institute of Technology (VIT), India. Previously, he held the position of a contract professor at the Department of Mechatronics Engineering, Jeju National University, South Korea. With more than a decade of expertise in teaching and research, Dr. Krishnamoorthy specializes in the synthesis and characterization of two-dimensional (2D) nanomaterials. His research is centered around Raman spectroscopic analysis of low-dimensional materials, the development of advanced energy harvesting and storage devices using 2D nanostructures beyond graphene, and the advancement of electrochemical energy storage (EES) systems for Electric Vehicles (EVs). Notably, Dr. Krishnamoorthy boasts a Hirsch index of 58, reflecting his significant contributions to the field, which include 6 patent applications and the publication of more than 125 research articles in prestigious international journals. His exceptional research achievements have earned him recognition as one of the top scientists (2%) in the world, spanning all scientific (nano) disciplines, as assessed by Stanford University, consistently for three consecutive years.

Contents

Preface **XIII**

Chapter 1 **1**
Optical Response of 3D Model Topological Nodal-Line Semimetal
by Sita Kandel, Godfrey Gumbs and Oleg L. Berman

Chapter 2 **17**
Graphene Nanosheets as Novel Nanofillers in an Epoxy Matrix for Improved Mechanical Properties
by Hadja Imane Beloufa, Mostapha Tarfaoui, Khalid Lafdi, Mohamed Daly and Amine Bendarma

Chapter 3 **37**
Unraveling the Role of Graphene Nanosheets in Electric Discharge Machining
by Kamlesh Paswan, Somnath Chattopadhyaya and Anil Dube

Chapter 4 **51**
Heavy Metals Adsorption by Nanosheet: Mechanism and Effective Parameters
by Mostafa Khosroupour Arabi and Morteza Ghorbanzadeh Ahangari

Chapter 5 **87**
Preparation of Siloxene-Graphene 2D/2D Heterostructures for High-Performance Supercapacitors in Electric Vehicles
by Karthikeyan Krishnamoorthy, Parthiban Pazhamalai, Rajavarman Swaminathan and Sang-Jae Kim

Chapter 6 **103**
Graphene Nanosheets for Fuel Cell Application
by Barsha Das, Sagnik Das, Soumyabrata Tewary, Sujoy Bose, Sandip Ghosh and Avijit Ghosh

Chapter 7 **121**
Biomedical Applications of Chitosan-Coated Nanosheets
by Sasireka Rajendran, Vinoth Rathinam, Vasanth Kumar, Manusree Kandasamy, Sharmila Selvi Muthuvel and Shanmugasundari Arumugam

Preface

Nanosheets are a class of material that consists of extremely thin, two-dimensional (2D) layers with nanoscale thickness and lateral size in the range of 100 to 1000 nm. After the discovery of graphene sheets, researchers explored a variety of 2D nanosheets from the family of transition metal dichalcogenides (TMDCs), MXenes, transition metal oxides/hydroxides, metalenes, siloxene, and so on. Nanosheets have gained significant attention in scientific research and various technological applications due to their unique properties such as large surface area, tunable properties, flexibility, mechanical strength, barrier properties, energy conversion, harvesting, and storage, catalysis, etc.

The chapters in this book are organized in a logical sequence, covering various aspects from the preparation to the characterization of nanomaterials. These aspects are explored in relation to specific applications, including optical responses, mechanical properties, electric discharge machining, heavy metal adsorption, fuel cells, supercapacitors for electric vehicles, and the biomedical sector. In addition to exploring the intrinsic properties of individual nanosheets, this text delves into the influence of nanosheet-based composites, their role as templates for nanocatalyst growth, and the significance of 2D/2D heterostructures in specific applications, offering a comprehensive examination. As the editor of this book, I would like to express my heartfelt appreciation to all the chapter authors for their invaluable contributions. This book serves as a valuable resource for scientists working in interdisciplinary science and technology, experts across various academic and industrial sectors, and students alike.

Dr. Karthikeyan Krishnamoorthy
Associate Professor,
Department of Physics,
School of Advanced Sciences (SAS),
Vellore Institute of Technology (VIT),
Vellore, Tamilnadu, India

Chapter 1

Optical Response of 3D Model Topological Nodal-Line Semimetal

Sita Kandel, Godfrey Gumbs and Oleg L. Berman

Abstract

We present a semi-analytical expression for both longitudinal and transverse optical conductivities of a model TNLSM employing the Kubo formula with emphasis on the optical spectral weight redistribution, deduced from appropriate Green's functions. In this semimetal, the conduction and valence bands cross each other along a one- dimensional curve protected by certain symmetry group in the 3D Brillouin zone. Although the crossing cannot be removed by any perturbations, it can be adjusted by continuous tuning of the Hamiltonian with a parameter α. When $\alpha > 0$, the two bands cross each other near the Γ point in the (k_x, k_y) plane of the first Brillouin zone making a nodal circle of radius $\sqrt{\alpha}$. The circle shrinks to point when $\alpha = 0$ and for $\alpha < 0$, the nodal circle vanishes and a gap opens around Γ. Numerical results for the longitudinal optical response of such TNLSM are investigated by varying the gap due to modifying α, the chemical potential μ, temperature T and the dephasing parameter η. The longitudinal optical conductivity is anisotropic along the direction parallel or perpendicular to the nodal ring. However, the transverse optical conductivity vanishes due to rotational symmetry.

Keywords: optical conductivity, nodal-line semimetal, topologically protected material, anisotropic energy band, linear response theory

1. Introduction

In recent times, topological semimetals (TSM) have come to the forefront of condensed matter physics following precise theoretical predictions, well controlled material synthesis and advance characterized techniques such as angle-resolved photoemission spectroscopy (ARPES), scanning tunneling microscopy (STM), and optical spectroscopy [1–4]. The ARPES directly probes the band structures of bulk and surface states, and information on the Fermi surface can also be revealed by STM through the quasiparticle interference process. Variety of TSM has been theoretically predicted and experimentally confirmed including Weyl [5], Dirac [6], and nodal-line semimetals [7], which correspond to different types of band crossings and the associated band topologies [8]. For these systems, the conduction and valence bands cross each other in the Brillouin zone (BZ). It is important to note that the crossing cannot be eliminated by perturbing the Hamiltonian without breaking either its crystalline or time-reversal symmetry. Therefore, all topological semimetals belong to symmetry protected topological phases of matter.

For Dirac semimetals, the conduction and valence bands have linear crossing at the Dirac point. If the time-reversal symmetry is broken, the Dirac point splits into two separated ones, and the system becomes a Weyl semimetal. On the other hand, for the nodal-line semimetal, the crossing of the valence and conduction bands forms a nodal ring (also called nodal line). The NLSM exhibit some peculiar properties which are absent in Dirac and Weyl semimetals. For example, the single nodal line in the NLSM can shrink to a point and vanish by continuously tuning the Hamiltonian in the absence of spin-orbital coupling (SOC). When strong SOC is added to the system, each nodal-line is either split or gapped due to hybridization between opposite spin components [9]. It is not true for the four bands crossing nodal-line for which the nodal-line is preserved even in the strong SOC regime and cannot be completely gapped out by tuning the Hamiltonian. In this work, we consider the two-bands crossing NLSM in the absence of SOC.

The surface states of topological materials are significantly affected by the number of layers and impurities. When the number of material layers is insufficient, the surface state will not be formed, or impurities in the system can readily destroy the surface states. In general, observation of surface states of topological materials such as drum-head surface states (DSS) of NLSM [2, 10–14] is still a great challenge. Our current work mainly focuses on the physical properties of bulk states.

When it came to be known that some topological semimetals exhibit different energy band dispersion relations than those of Dirac and Weyl materials, and making a nodal ring at the cross-section of the valence and conduction bands, NLSM have offered an attractive research platform for exploring a variety of novel phenomena. Various materials have been theoretically predicted or experimentally confirmed as NLSM, including ZrXY (X = Si, Ge, Y = S, Se, Te) [3, 15, 16], graphene networks [14], compressed black phosphorus [17] as well as $HgCr_2Se_4$ [18]. The optical conductivity of monolayer and bilayer graphene, and few-layer epitaxial graphite has been reported recently [19]. These studies have yielded useful information regarding the electron dynamics. We refer to reference [20] for a review of other remarkable properties of these TSM systems as well as a discussion of possible technological applications. Here, we consider the optical conductivity of a NLSM model system with special emphasis on the optical spectral weight redistribution due to changes in the chemical potential caused by charging as well as an adjustable parameter α which governs the degree of crossing between the valence and conduction bands. The parameter α can be manipulated by strain. The effect of mechanical strain on the optical properties of nodal-line semimetal ZrSiS has been studied using first-principles calculations in Ref. [21]. According to reference [21], frequency-independent optical conductivity is robust with respect to uniaxial compressive strain of up to 10 GPa.

Similar studies have reported previously. In 2016/17, Carbotte and his team [22, 23] studied the optical response of 3D NLSM and came to the conclusion that it presents two-dimensional (2D) Dirac-like response in the low-photon energy regime and 3D Dirac-like response in the high-energy photon limit. Later on in 2017, Sashin Barati [24] also showed that the optical conductivity of NLSM is anisotropic. Both those authors considered different toy models in order to explain NLSM with different Fermi surfaces. However, none of these results totally agree with the paper by Habe and Koshino [25]. Koshino and his team studied theoretically the dynamical conductivity of ZrSiS by using a multi-orbital tight-binding model based on a first-principles band calculation. According to Habe and Koshino, ZrSiS type NLSM attributes the optical conductivity which is neither liner like 2D- nor like 3D. The interband contribution first increases slowly for some frequency range, then it is decreased to smaller value and saturates for large frequency. This motivates us to study the optical response of similar materials using a simple model Hamiltonian and the Kubo formula which yields semi-analytic results. We successfully presented semi-analytical formula for both transverse and longitudinal optical conductivity including both intra and interband contributions of the model Hamiltonians. Our result is interesting because the numerical solutions closely resemble Habe and Koshino's results and are neither like 2D nor 3D Dirac or Weyl semimetal. The transverse optical conductivity of NLSM is found to vanish due to rotational symmetry along the ring of NLSM. This result is also supported by a recent study of Wang [26], who demonstrated the anomalous Hall optical conductivity in tilted topological nodal-line semimetals.

The rest of this work is organized as follows. In section II, we present the theoretical formalism for the calculation of the optical conductivity based on the simplest nearest-neighbor Hamiltonian. This model involves an adjustable parameter α associated with a nodal circle which may shrink to a point when $\alpha = 0$. Here, we have managed to derive a conveniently simple semi-analytical expression for the frequency-dependent conductivity for arbitrary chemical potential, making this a convenient result for experimentalists. In section III, we present numerical results demonstrating the anisotropy of the system, the α-dependence and charging pertaining to the biasing of the NLSM sample. Section IV is devoted to a summary of our results.

2. Theoretical background

The NLSM is modeled by a slab consisting of several nanosheets of single atoms along the z axis. We apply periodic boundary conditions along the z axis since we are only interested in bulk properties of the NLSM. This is shown schematically in **Figure 1**. Consequently, in this work, only bulk states are included in our calculations and not drum head surface states which arise for a slab of finite thickness. Such topologically protected DSS can be exhibited in a slab geometry with suitable thickness, e.g., with sixty unit cells for Ca_3P_2 [27].

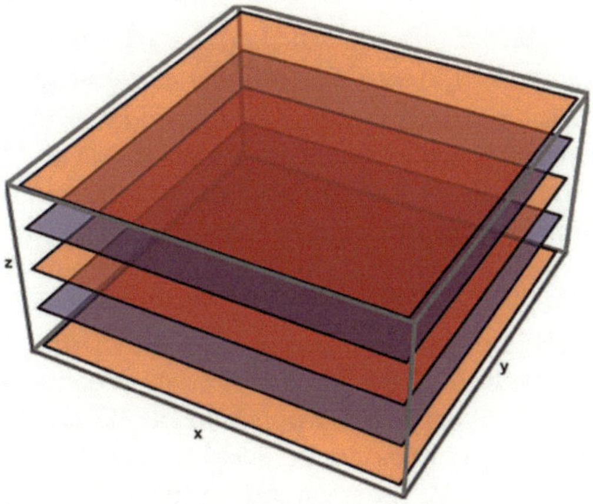

Figure 1.
Schematic representation of a nodal-line semimetal consisting of several nanosheets of atoms stacked along the z axis making up a slab (color online).

Let us consider a Hamiltonian for which the nodal-line is stable against perturbations. However, the nodal-line may still shrink continuously to a point. In this regard. We turn to a single-spin effective Hamiltonian of a material such as ZrSiSe and ZrSiTe [3, 4]. In Ref. [2], the band structure of ZrSiTe was obtained using *ab initio* calculations. The idea here is to carry out a calculation based on a model Hamiltonian which mimics the essential features of the true band structure near the Γ point of a NLSM such as ZrSiTe. In general, these band structures obtained from first-principles calculations or the generalized tight-binding model are very complicated. Of course, not all features are reproduced by this model Hamiltonian throughout the Brillouin zone but it could serve as a useful tool for generating qualitative results to get a better understanding of this growing class of topological semimetals. For the nodal line, enclosed with either a ring or a sphere, the symmetry group Z_2 is the topological classification of the wave functions on the ring/sphere [20]. Therefore, a nodal ring is characterized by two independent Z_2 indices, denoted by ζ_1 and ζ_2, defined on a ring that links with the line and on a sphere that encloses the whole line. According to reference [20], all topological nodal rings, protected by this symmetry group with respect to an arbitrarily small perturbation, are characterized by $\zeta_1 = 1$. For NLSMs with $\zeta_1 = 1$ and $\zeta_2 = 0$, our toy model Hamiltonian is given by [20]

$$\hat{H} = \left(\alpha - k^2\right)\hat{\sigma}_z + k_z\hat{\sigma}_x, \tag{1}$$

where energy is measured in units of $\hbar v_F k_F$ with v_F the Fermi velocity equal to 10^6m/s and k_F is the Fermi wave vector. Also, the wave vector $\mathbf{k} = \left(k_x, k_y, k_z\right)$ is scaled in terms of π/a where a is a lattice constant and $-1 \le k_i \le 1$ with $i = x, y, z$ in order to ensure that the model is restricted near the Γ point. For convenience, we will refer to this hypercubic region as the Brillouin zone. We have $\hat{\sigma}_x, \hat{\sigma}_z$ representing Pauli matrices. The quantity α is an adjustable parameter which can be employed to vary the energy gap around the Γ point for which the energy bands are given by

$$\varepsilon_s(\mathbf{k}) = s\sqrt{k_z^2 + \left(\alpha - k^2\right)^2}. \tag{2}$$

The band degeneracy occurs as $k_x^2 + k_y^2 = \alpha$ and $k_z = 0$. The two bands touch at $\mathbf{k} = 0$ when $\alpha = 0$. If $\alpha > 0$, the two subbands cross each other near the Γ point on the $\left(k_x - k_y\right)$ plane for fixed $k_z = 0$, describing a nodal circle of radius $\sqrt{\alpha}$ which shrinks to a point at $\mathbf{k} = 0$ as α is decreased to zero and the nodal circle vanishes and the valance and conduction bands are gapped out when $\alpha < 0$. The 3D energy dispersion is presented in **Figure 2**. In **Figure 3**, we plot the energy bands in the first BZ hypercube along lines joining points of high symmetry, corresponding to $\alpha = 0$ and $\alpha = 0.5$. These two contrasting cases demonstrate how the band gap near Γ may be tuned by varying this parameter which may be manipulated as a function of strain. It is clear from these results that the band structure is modified by finite α throughout the BZ. **Figure 3(b)** shows that the bulk band gap closes at an even number of discrete points near the Γ point. These special gap closing points in the BZ are protected by crystalline symmetry and play a role in the behavior of Weyl semimetals [28]. The detailed analysis of the topology of NLSMs, corresponding to the Hamiltonian in Eq. (1), is provided in [20]. This modification of the band structure has significant effects on the physical properties we calculate in this paper and may also affect the thermoelectric and Boltzmann transport as well as the plasmon excitations.

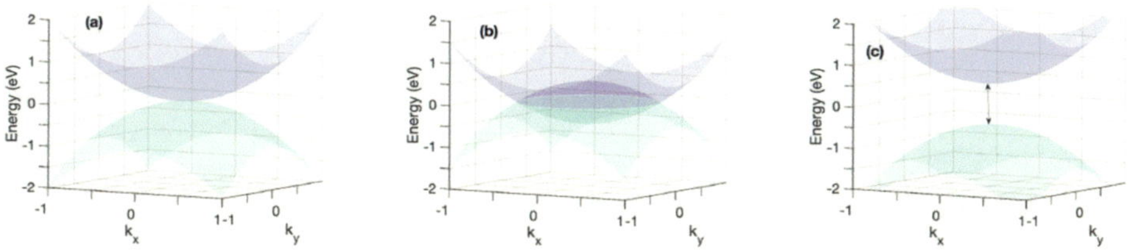

Figure 2.
(color online) the 3D contour energy bands of the NLSM described by the model Hamiltonian in Eq. (1) on the $k_z = 0$ plane for (a) $\alpha = 0$, (b) $\alpha = 0.5$ and (c) $\alpha = -0.5$. For positive α, two bands cross each other making a nodal circle of radius $\sqrt{\alpha}$. For α equal to zero, the nodal circle shrink to a point and for $\alpha < 0$, the nodal circle vanishes and the bands are gapped out.

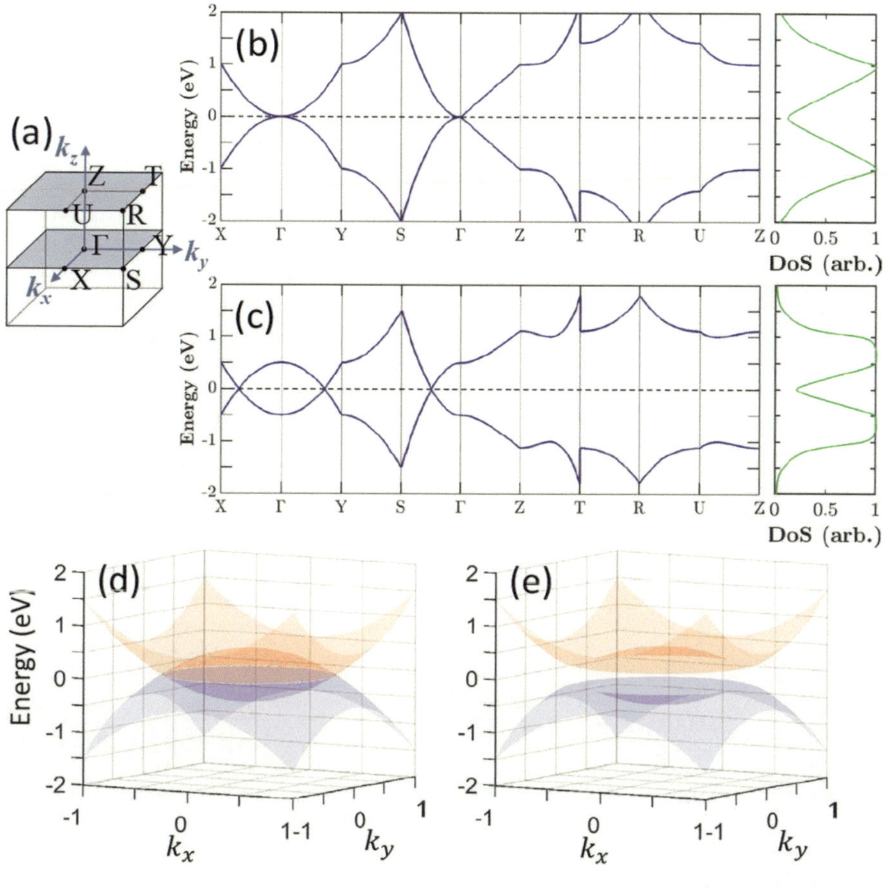

Figure 3.
(color online) (a) shows the 3D hypercube, or reduced Brillouin zone, with its high-symmetry points for bulk NLSM. Bulk band structures between the high symmetry points (left) and corresponding density of states (right) for (b) $\alpha = 0$ and (c) $\alpha = 0.5$. The 3D contour bands are shown at low energy on the selected (k_x, k_y) plane for (d) $k_z = 0$ and (e) $k_z = 0.2$.

The normalized eigenvectors φ_s (for $s = \pm$) of the Hamiltonian (1) are given by

$$\varphi_s(\mathbf{k}) = \begin{pmatrix} \beta_s(\mathbf{k}) \\ \eta_s(\mathbf{k}) \end{pmatrix} \frac{e^{i\mathbf{k}\cdot\mathbf{r}}}{\sqrt{\mathcal{V}}}, \tag{3}$$

where $\mathcal{V}$ is a normalization volume. and β_s and η_s are given by

$$\beta_s^2(\mathbf{k}) = \frac{k_z^2}{k_z^2 + \left(k^2 - \alpha + s\sqrt{(k^2-\alpha)^2 + k_z^2}\right)^2}; \tag{4}$$

$$\eta_s^2(\mathbf{k}) = \frac{\left(k^2 - \alpha + s\sqrt{(k^2-\alpha)^2 + k_z^2}\right)^2}{k_z^2 + \left(k^2 - \alpha + s\sqrt{(k^2-\alpha)^2 + k_z^2}\right)^2}, \tag{5}$$

where β_s and η_s satisfy the normalization condition: $\beta_s^2 + \eta_s^2 = 1$.
For the density-of-states, we have

$$N(\omega) = N_f \int_{BZ} \frac{d^3\mathbf{k}}{(2\pi)^3} \sum_{s=\pm} \delta(\omega - \varepsilon_s(\mathbf{k})) \ . \tag{6}$$

We now introduce the Green's function defined by $\hat{\mathcal{G}}^{-1}(z) = z\hat{I} - \hat{H}$ which in matrix form is written as

$$\hat{\mathcal{G}}(\omega, \mathbf{k}) = \frac{1}{\mathcal{D}(\omega, \mathbf{k})} \begin{pmatrix} \omega + (\alpha - k^2) & k_z \\ k_z & \omega - (\alpha - k^2) \end{pmatrix}, \tag{7}$$

with $\mathcal{D}(\omega, \mathbf{k}) \equiv \omega^2 - (\alpha - k^2)^2 - k_z^2 = (\omega - \varepsilon_+(\mathbf{k}))(\omega - \varepsilon_-(\mathbf{k}))$.
Now, the spectral function representation of the Green's function is for $i,j = 1,2$

$$\mathcal{G}_{ij}(z) = \int_{-\infty}^{\infty} \frac{d\omega'}{2\pi} \frac{A_{ij}(\omega')}{z - \omega'}. \tag{8}$$

For $i,j = 1$,

$$G_{11}(z) = \int_{-\infty}^{\infty} \frac{d\omega' A_{11}(\omega')}{2\pi(z - \omega')} = \frac{z + (\alpha - k^2)}{(z - \varepsilon_+)(z - \varepsilon_-)} \tag{9}$$

Resolving the right-hand side of this equation into partial fractions, we obtain

$$\frac{1}{(\varepsilon_+ - \varepsilon_-)} \left[\frac{[\varepsilon_+ + (\alpha - k^2)]}{(z - \varepsilon_+)} - \frac{[\varepsilon_- + (\alpha - k^2)]}{(z - \varepsilon_-)} \right] = \int_{-\infty}^{\infty} \frac{d\omega' A_{11}(\omega')}{2\pi(z - \omega')} = G_{11}(\omega). \tag{10}$$

From this, we deduce

$$A_{11}(\mathbf{k}, \omega) = \frac{2\pi}{(\varepsilon_+ - \varepsilon_-)} \left[[\varepsilon_+ + (\alpha - k^2)] \ \delta(\omega - \varepsilon_+) - [\varepsilon_- + (\alpha - k^2)] \ \delta(\omega - \varepsilon_-) \right]. \tag{11}$$

Similarly, we can obtain other spectral functions as follows.

$$
\begin{aligned}
A_{22}(\mathbf{k},\omega) &= \frac{2\pi}{\varepsilon_+ - \varepsilon_-}\left\{\left(\varepsilon_+ - (\alpha - k^2)\right)\delta(\omega - \varepsilon_+) - \left(\varepsilon_- - (\alpha - k^2)\right)\delta(\omega - \varepsilon_-)\right\},\\
A_{12}(\mathbf{k},\omega) &= A_{21}(\mathbf{k},\omega) = \frac{2\pi k_z}{\varepsilon_+ - \varepsilon_-}\{\delta(\omega - \varepsilon_+) - \delta(\omega - \varepsilon_-)\}.
\end{aligned}
\tag{12}
$$

The real (absorptive) part of the optical conductivity can be written in the form

$$
\begin{aligned}
\sigma_{\alpha\beta}(\Omega) = \frac{N_f e^2}{2\Omega}\int_{-\infty}^{\infty}\frac{d\omega}{2\pi}[f(\omega - \mu) - f(\omega + \Omega - \mu)]\\
\times \int_{BZ}\frac{d^3\mathbf{k}}{(2\pi)^3} Tr\left[\hat{v}_\alpha \hat{A}(\mathbf{k},\omega + \Omega)\hat{v}_\beta \hat{A}(\mathbf{k},\omega)\right],
\end{aligned}
\tag{13}
$$

where N_f is a degeneracy factor for the spin and at temperature T, we have $f(x) = 1/[\exp(x/T) + 1]$ as the Fermi function and μ is the chemical potential. Also, α, β represent spatial coordinates x, y, z. In doing our numerical calculations, we choose the zero temperature case T = 0 K, and the finite temperature case T = 300 K. For the longitudinal in-plane conductivity, $\hat{\sigma}_{xx}(\Omega)$, we have $\hat{v}_\alpha = \hat{v}_\beta = \hat{v}_x = -2k_x\hat{\sigma}_z$. Putting these results together, we conclude that the longitudinal conductivity $\sigma_{xx}(\Omega)$ is

$$
\sigma_{xx}(\Omega) = \frac{N_f e^2}{8\hbar\Omega\pi^4}\int_{|\mu|-\Omega}^{|\mu|} d\omega \int_{BZ} d^3\mathbf{k}\ k_x^2\ \gamma(\mathbf{k},\omega;\Omega), \tag{14}
$$

where

$$
\begin{aligned}
\gamma(\mathbf{k},\omega;\Omega) \equiv 4\pi^2\Bigg\{&\left[1 - \frac{k_z^2}{E_k^2}\right][\delta(\omega + \Omega - E_k)\delta(\omega - E_k) + \delta(\omega + \Omega + E_k)\delta(\omega + E_k)]\\
&+ \frac{k_z^2}{E_k^2}[\delta(\omega + \Omega - E_k)\delta(\omega + E_k) + \delta(\omega + \Omega + E_k)\delta(\omega - E_k)]\Bigg\}.
\end{aligned}
\tag{15}
$$

For temperature $T = 0$ K, the Fermi function acts as a step function so the integration limit changes to $|\mu| - \Omega$ to $|\mu|$. In order to simplify the expressions, we chose E_k to represent the magnitude of energy eigenvalue. Hence, we replaced $\varepsilon_+(\mathbf{k}) = E_k$ and $\varepsilon_-(\mathbf{k}) = -E_k$. These terms give the intraband and interband contributions to the conductivity due to transitions within the conduction band and from transitions between the valence and conduction bands, respectively. The significance of these contributions is determined by the level of doping, the frequency Ω and implicitly by α through the energy dispersion $\varepsilon_s(\mathbf{k})$. The limit when $\Omega \to 0$ corresponds to the Drude conductivity [29]. We analyzed this intraband (Drude) conductivity using a Lorentzian representation for the δ-function. Additionally, the terms proportional to $1 - k_z^2/E^2$ in Eq. (15) vanish when $\Omega > 0$ and gives only the intraband conductivity.

Similarly, for $\alpha = \beta = y$, we have $\hat{v}_\alpha = \hat{v}_\beta = \hat{v}_y = -2k_y\hat{\sigma}_z$. The expression for $\sigma_{yy}(\Omega)$ comes out to be

$$\sigma_{yy}(\Omega) = \frac{N_f e^2}{2\Omega} \int_{-\infty}^{\infty} \frac{d\omega}{2\pi} \left[f(\omega - \mu) - f(\omega + \Omega - \mu)\right] \int_{BZ} \frac{d^3\mathbf{k}}{(2\pi)^3} \left(2k_y\right)^2 \gamma(\mathbf{k}, \omega; \Omega). \quad (16)$$

Additionally, for $\alpha = \beta = z$, we have $\hat{v}_\alpha = \hat{v}_\beta = \hat{v}_z = -2k_z\hat{\sigma}_z + \hat{\sigma}_x$. The $\sigma_{zz}(\Omega)$ is

$$\sigma_{zz}(\Omega) = \frac{N_f e^2}{2\Omega} \int_{-\infty}^{\infty} \frac{d\omega}{2\pi} \left[f(\omega - \mu) - f(\omega + \Omega - \mu)\right] \int_{BZ} \frac{d^3k}{(2\pi)^3} \left\{(2k_z)^2 \gamma(\mathbf{k}, \omega; \Omega) - 4k_z \kappa(\mathbf{k}, \omega; \Omega) + \zeta(\mathbf{k}, \omega; \Omega)\right\}, \quad (17)$$

where

$$\kappa(\mathbf{k}, \omega; \Omega) \equiv \frac{4\pi^2 k_z \left(\alpha - k^2\right)}{E_k^2} \Big\{\delta(\omega + \Omega - E_k)\delta(\omega - E_k) - \delta(\omega + \Omega + E_k)\delta(\omega - E_k)$$
$$- \delta(\omega + \Omega - E_k)\delta(\omega + E_k) + \delta(\omega + \Omega + E_k)\delta(\omega + E_k)\Big\}, \quad (18)$$

$$\zeta(\mathbf{k}, \omega; \Omega) \equiv 4\pi^2 \left\{ \left(1 - \frac{k_z^2}{E_k^2}\right) \left[\delta(\omega + \Omega - E_k)\delta(\omega + E_k) + \delta(\omega + \Omega + E_k)\delta(\omega - E_k)\right] \right.$$
$$\left. + \frac{k_z^2}{E_k^2} \left[\delta(\omega + \Omega - E_k)\delta(\omega - E_k) + \delta(\omega + \Omega + E_k)\delta(\omega + E_k)\right] \right\}. \quad (19)$$

All three functions γ, κ and ζ are even functions of $\mathbf{k} = \left(k_x, k_y, k_z\right)$.

For the transverse component $\sigma_{xy}(\Omega)$, we must set $\alpha = x$, $\beta = y$, $\hat{v}_\alpha = \hat{v}_x = -2k_x\hat{\sigma}_z$ and $\hat{v}_\beta = \hat{v}_y = -2k_y\hat{\sigma}_z$ Then we obtain

$$\sigma_{xy}(\Omega) = \frac{N_f e^2}{2\Omega} \int_{-\infty}^{\infty} \frac{d\omega}{2\pi} \left[f(\omega - \mu) - f(\omega + \Omega - \mu)\right] \int \frac{d^3\mathbf{k}}{(2\pi)^3} \left(4k_x k_y\right) \gamma(\mathbf{k}, \omega; \Omega). \quad (20)$$

Since the integrand of σ_{xy} is an odd function of k_x, the transverse component of the conductivity along $k_x k_y$ is zero. Similarly, for $\sigma_{yz}(\Omega)$, setting $\alpha = y$, $\beta = z$, $\hat{v}_\alpha = \hat{v}_y = -2k_y\hat{\sigma}_z$ and $\hat{v}_\beta = \hat{v}_z = -2k_z\hat{\sigma}_z + \hat{\sigma}_x$ and for $\sigma_{xz}(\Omega)$, setting $\alpha = x$, $\beta = z$, $\hat{v}_\alpha = \hat{v}_x = -2k_x\hat{\sigma}_z$ and $\hat{v}_\beta = \hat{v}_z = -2k_z\hat{\sigma}_z + \hat{\sigma}_x$ we arrive at

$$\sigma_{yz}(\Omega) = \frac{N_f e^2}{2\hbar\Omega} \int_{-\infty}^{\infty} \frac{d\omega}{2\pi} \left[f(\omega - \mu) - f(\omega + \Omega - \mu)\right] \int_{BZ} \frac{d^3\mathbf{k}}{(2\pi)^3} \left\{-2k_y\ \kappa(\mathbf{k}, \omega; \Omega) + 4k_y k_z\ \gamma(\mathbf{k}, \omega; \Omega)\right\}, \quad (21)$$

$$\sigma_{xz}(\Omega) = \frac{N_f e^2}{2\hbar\Omega} \int_{-\infty}^{\infty} \frac{d\omega}{2\pi} \left[f(\omega - \mu) - f(\omega + \Omega - \mu)\right] \int_{BZ} \frac{d^3\mathbf{k}}{(2\pi)^3} \left\{-2k_x\ \kappa(\mathbf{k}, \omega; \Omega) + 4k_x k_z\ \gamma(\mathbf{k}, \omega; \Omega)\right. . \quad (22)$$

Clearly, the integrands in Eq. (21) and Eq. (22) are both odd functions of k_x and k_y thereby making $\sigma_{yz}(\Omega)$ and $\sigma_{xz}(\Omega)$ zero. We now employ these results in the following section to carry out our numerical calculations for the longitudinal optical conductivity. These results exhibit the anisotropy of this NLSM system and demonstrate their dependence on the parameter α and the chemical potential μ as the frequency is varied.

3. Results and discussion for the longitudinal conductivity

We now examine in detail with the help of our numerical solutions the absorptive part of the longitudinal optical conductivity of an anisotropic nodal-line semimetal along the radial x, y and axial z directions. For these calculations, we have used $N_f = 2$ corresponds to spin degeneracy, the small momentum cutoff with $-1 \leq k_{x,y,z} \leq 1$ to lie within a hypercube or so-called first Brillouin zone and a Lorentzian representation for the Dirac δ-function, i.e., $\delta(x) \to \frac{\eta}{\pi} \frac{1}{\eta^2 + x^2}$ with a broadening parameter η. This broadening takes into account the scattering due to non-magnetic impurities and lattice defects. The broadening is manifested in the optical conductivity as an effective transport scattering rate of $\frac{1}{\tau} = 2\eta$. In **Figure 4**, the effect of the broadening parameter η on the optical conductivity has been illustrated. All of the presented results are calculated for $\eta = 0.5$. It is convenient to scale the conductivity by $\sigma_0 = e^2/4\hbar$ and we set $\hbar = 1$ in the above Kubo formula. The finite temperature responses are also analyzed using the Fermi-Dirac distribution function at temperature T by

$$f(\xi) = \frac{1}{1 + \exp(\xi - \mu)/k_B T} = \frac{1}{2}\left\{1 - \tanh\left(\frac{\beta(\xi - \mu)}{2}\right)\right\}, \tag{23}$$

where $\beta = 1/k_B T$.

The band structure of bulk NLSM along the high symmetry directions in the cubic Brillouin zone shows that the energy bands are anisotropic along the three perpendicular axes. When we go from Γ to X and Y in **Figure 3**, the energy bands are symmetric but it is not so along the z axis. For $\alpha = 0$, the two subbands touch at Γ making a flat dispersionless band in a small region around it. It is clearly seen that the dispersion around Γ is not linear like that for Dirac or Weyl semimetal. Therefore, we should not expect Dirac or Weyl-like optical conductivity even though two bands just touch at a point. The transverse optical conductivity vanishes because of rotational symmetry of the NLSM. The longitudinal conductivities along different directions are obtained

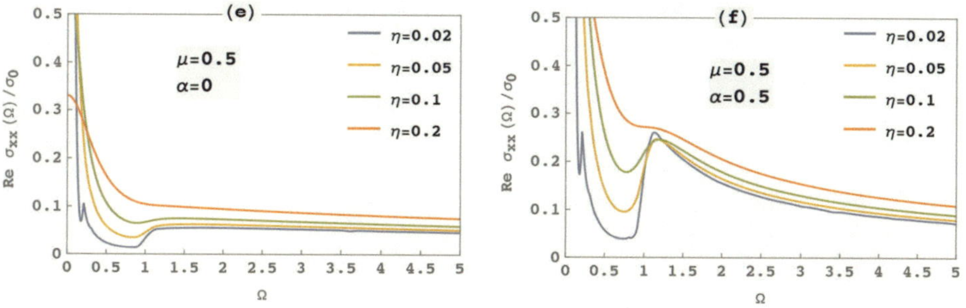

Figure 4.
(color online) real part of the longitudinal optical conductivity at T = 0 K along x axis measured for $\mu = 0.5$ and for $\alpha = 0$ in (e) and for $\alpha = 0.5$ in (f). The different curves correspond to chosen dephasing parameter η. In the Lorentzian representation of the delta function, η tends to zero. We have used η equal to 0.5 for numerical calculation.

accordingly with the band structure and density of states plots. The anisotropic optical responses of the NLSM are depicted in **Figure 5** considering separately two chosen values of μ. The conductivities along x and y are exactly the same. However, along the z direction, the interband contribution for both pristine ($\mu = 0$) and the dopped ($\mu = 0.5$) NLSM has greater value. The delta peak near $\Omega = 0$ is called intraband or Drude conductivity. This peak is formed due to the transition of electrons within the conduction band which we have numerically calculated using the Lorentzian representation for the Dirac δ-function in the expression. For the numerical calculations, we have considered only two low energy bands, and hence the effect of higher energy bands is neglected. The density of states has peak values for some energy range. The conductivity is expected to be minimal for energy greater than that range. The 3D contour bands at low energy in the k_x,k_y plane are either crossed or gapped depending on the value of k_z. When $k_z = 0$, two bands cross. However, they open a gap on another plane for $k_z = 0.2$. While integrating over the hypercube, the effects from both planes are included which leads to finite values for both interband and intraband conductivity even for pristine NLSM.

In the case of finite μ, we have zero conductivity for frequency less than 2μ and a finite jump for frequency greater than 2μ which reveals that the conductivity is due to interband transitions and that transitions obey the Pauli exclusion principle. For photon energy greater than 2μ, the conductivity reaches a maximum value and then is decreased to a minimum saturated value. The peak of the curve depends significantly on the tunable parameter α. The conductivity becomes flat when α goes from positive to negative value. Specifically, the optical conductivity is reduced when a sizable gap is opened by tuning the value of α. Carbotte [22] and Mukherjee [23] have reported the flat conductivity in NLSM with notable finite height. However, our result is more

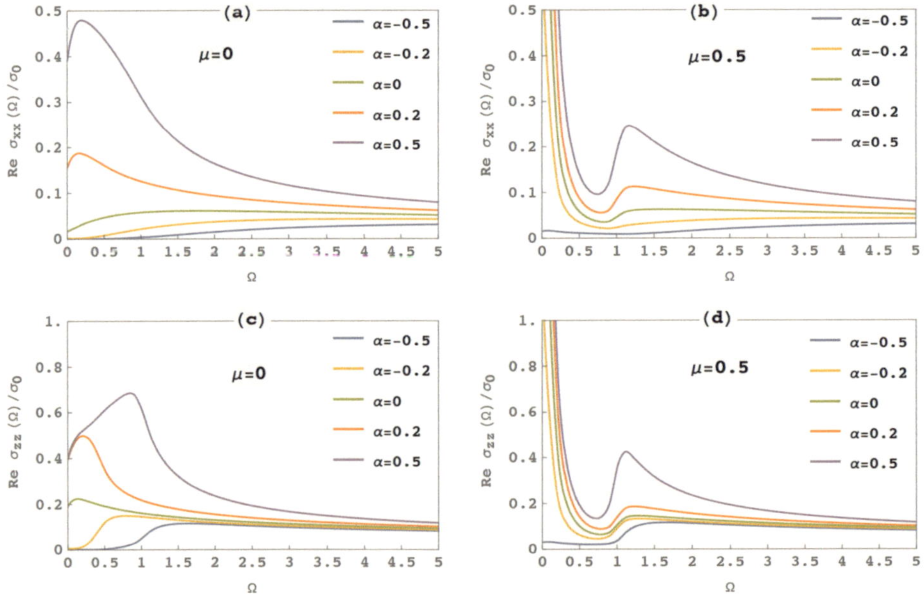

Figure 5.
(color online) real part of the longitudinal optical conductivity at T = 0 K measured along the x and z directions in the top to bottom panels, respectively, with (a) and (d) $\mu = 0$ and (b) and (c) $\mu = 0.5$ for different chosen values of the tuning parameter α. As the value of α is decreased, the conductivity becomes flatter and less dependent on frequency. The given NLSM admits similar responses along x and y and slightly different response along the z direction both quantitatively and qualitatively.

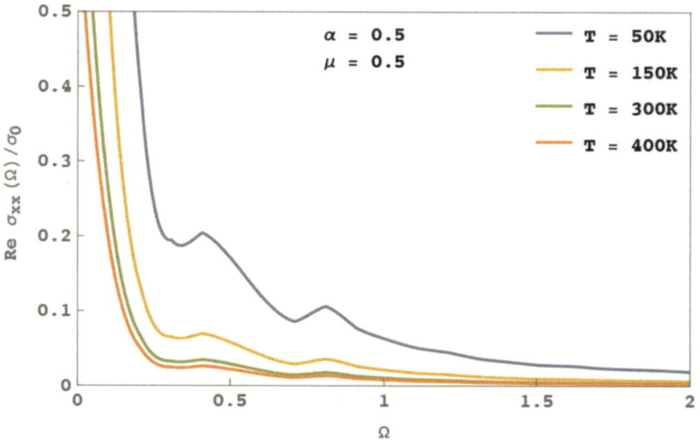

Figure 6.
(color online) finite temperature response of the optical conductivity measured in the x direction for doped NLSM at $\alpha = 0.5$. When the temperature is increased from 0 K, the conductivity reduced at low frequency region.

in line with that for ZrSiS [25]. Habe and Koshino illustrated that the optical conductivity of ZrSiS-like NLSM acquires the peak value and then sharply decreases to a flat plateau region and again rises up in the high frequency region. Throughout these calculations, we have considered a toy-model Hamiltonian with only two subbands and small momentum cutoff. Consequently, these results are applicable for small values of chemical potential and for low frequency. For large values of chemical potential and for higher frequency range, the contribution of other subbands to the conductivity may be substantially significant and the conductivity may again increase. The effect due to finite temperature manifests itself through similar qualitative behavior. However, this is quantitatively less responsive compared to that at 0 K. **Figure 6** demonstrates that when the temperature is increased from 0 K, the interband optical conductivity is reduced. This is because the thermally excited electrons and holes perturb the optical transition from the valance to the conduction band in the low-frequency regime. Furthermore, when the temperature is increased from room temperature, the separation between the conductivity curves is drastically reduced. This means that the effect due to temperature is more significant in the low-temperature regime.

4. Summary and conclusions

In conclusion, we have theoretically investigated the optical responses of nodal-line semimetals using a simple model in conjunction with the Kubo formula for the conductivity. We have derived closed-form semi-analytical expressions for the longitudinal components of the optical conductivity for this simple model whose energy band structure is anisotropic in **k**-space. Although the model Hamiltonian is simple, the energy band structure still bears similarities to calculated results using the tight-binding model near the Fermi level of a pristine nodal-line semimetal. The effect due to strain is introduced through a parameter α in the model Hamiltonian, thereby making this simple model useful for making qualitative predictions of the NLSM using this heuristic approach. This is bolstered not only by the similarities in the band structure but by the overall qualitative behavior of the conductivity when compared with ZrSiS, for example. We have presented numerical results for the optical conductivity as a function of incident photon energy at both zero and finite temperatures. We analyzed the results considering the effect of strain for the simple model of nodal-line semimetal and chemical potential. We have plotted the band structure along lines joining high-symmetry points in a reduced hypercubic region in (k_x, k_y, k_z)-space. We emphasize that the model Hamiltonian is particularly representative of the low-energy bands near the Γ point.

The transverse conductivity vanishes due to rotational symmetry around the nodal ring. The Drude conductivity in the limit $\Omega \to 0$ is the result of impurity scattering which is phenomenologically introduced through a Lorentzian representation of the Dirac δ-function with a weak dephasing factor. For charge neutral/pristine NLSM, the interband conductivity is significant near zero frequency. However, for doped NLSM with chemical potential $\mu \neq 0$, the interband conductivity rises up to the peak value at $\Omega = 2\mu$ and then decreases. For frequency less than 2μ, Pauli blocking makes electrons unable to make transitions between the valence and conduction bands. At higher photon energies, the transition saturates at its minimum value. This can be explained as a consequence of the cubic Brillouin zone resulting in the value of our wave vector $\mathbf{K} = (k_x, k_y, k_z)$ is restricted to $-1k_i \leq 1$ $(i = x, y, z)$, scaled by $\frac{\pi}{a}$ where a is a lattice constant. The Hamiltonian $\hat{H}$, Ω and μ are all scaled in terms of the energy unit $\hbar v_F \pi / a$. Therefore, the conductivity of NLSM depends on the Fermi velocity v_F. The height and width of the peak depend principally on the tuning parameter (α). For $\alpha = 0$, the curve is almost flat at very small transition. At $\alpha = 0$, the bands touch at a point but the quadratic dispersion near Γ is not linear like Dirac or Weyl. Therefore, the conductivity is not flat like that of 2D and 3D Dirac materials as explained in [22–24]. As we increase the value of α from 0 to larger values, the peak becomes sharper and sharper. In any event, they all converge to a minimum saturated value in the higher frequency region. This is also in agreement with the density of states plot. The density of states has a peak value at a certain energy and zero over a range which could be adjusted by manipulating the value of α. The anisotropic dispersion relations along radial (k_x, k_y) and axial (k_z) directions result in their anisotropic conductivity along these directions. The conductivity along x and y is identical yet the conductivity along z deviates slightly both quantitatively and qualitatively. The anisotropic conductivities in Shahin's [24] paper are explained as a consequence of an anisotropic Fermi surface where they have considered the toroidal shape. These properties indicate possible applications involving microelectronics technologies, integrated circuits and devices, ultrafast modulators and high performance transistors.

We have also calculated the finite temperature-dependent response of the interband conductivity. Although the temperature suppresses transitions between subbands and reduces the conductivity, there is still qualitative similarity between the behavior of the response for the model Hamiltonian adapted here and that derived with the use of the tight-binding model. Therefore, the overall response of the optical conductivity for the chosen model Hamiltonian is suitable for investigating NLSM rather than that of Dirac or Weyl semimetals. The cubic Brillouin zone and the band structure plots suggest that the dispersion relations of our model Hamiltonian is more or less appropriate for drawing comparisons with that of ZrSiS. As a matter of fact, a few years ago, the dynamical optical conductivity of ZrSiS was investigated using a multi-orbital tight-binding model based on the first-principles band calculation [25] which agrees with our results that the optical conductivity of some TNLSM are significantly divergent from Dirac and Weyl semimetals. The reduced conductivity in the high frequency region was not fully explored in previous similar studies. While analyzing our results, we should not forget that we have considered the low energy, low-band model Hamiltonian. Therefore, the results would not be reliable for very high chemical potential and very high photon energies. In the higher frequency region, the effect of other bands and the higher order terms in the wave vector may not be negligible. This may result in substantial effect on the obtained optical conductivity. Finally, we came to the conclusion that different nodal-line semimetals described by specific model Hamiltonians may not have the same optical response in chosen frequency regions. The model Hamiltonian we considered has ZrSiS like dispersion relation and the conductivity also mimics well qualitatively with the overall behavior and this is encouraging to investigate the self-sustained charge density oscillations, i.e., plasmons.

Acknowledgements

G.G. would like to acknowledge the support from the Air Force Research Laboratory (AFRL) through Grant No. FA9453-21-1-0046. We would like to thank Dr. Po-Hsin Shih for helpful discussions and guidance on the numerical calculations and interpretation of the results.

Author details

Sita Kandel[1,2], Godfrey Gumbs[1,2,3]* and Oleg L. Berman[2,4]

1 Department of Physics, Hunter College, City University of New York, New York, USA

2 The Graduate School and University Center, The City University of New York, New York, USA

3 Donostia International Physics Center (DIPC), San Sebastian, Basque Country, Spain

4 Physics Department, New York City College of Technology, The City University of New York, Brooklyn, New York, USA

*Address all correspondence to: ggumbs@hunter.cuny.edu

References

[1] Lv BQ, Qian T, Ding H. Experimental perspective on three-dimensional topological semimetals. Reviews of Modern Physics. 2021;**93**:025002

[2] Muechler L, Topp A, Queiroz R, Krivenkov M, Varykhalov A, Cano J, et al. Modular arithmetic with nodal lines: Drumhead surface states in ZrSiTe. Physical Review X. 2020;**10**:011026

[3] Pan H, Tong B, Jihai Y, Wang J, Dongzhi F, Zhang S, et al. Three-dimensional anisotropic magnetoresistance in the Dirac node-line material ZrSiSe. Scientific Reports. 2018; **8**:9340. DOI: 10.1038/s41598-018-27148-z

[4] Yinming Shao AN, Rudenko JH, Sun Z, Zhu Y, Seongphill Moon AJ, Millis SY, et al. Electronic correlations in nodal-line semimetals. Nature Physics. 2020;**16**:636

[5] Murakami S. Phase transition between the quantum spin hall and insulator phases in 3D: Emergence of a topological gapless phase. New Journal of Physics. 2007;**9**:356

[6] Wang Z, Sun Y, Chen X-Q, et al. Dirac semimetal and topological phase transitions in a_3 Bi (a=N, K, Rb). Physical Review B. 2012;**85**:195320

[7] Burkov A, Hook M, Balents L. Topological nodal semimetals. Physical Review B. 2011;**84**:235126

[8] Yang M-X, Luo W, Chen W. Quantum transport in topological nodal-line semimetals. Advances in Physics X. 2022;7:2065216

[9] Fang C, Chen Y, Kee H-Y, Fu L. Topological nodal line semimetals with and without spin-orbital coupling. Physical Review B. 2015;**92**: 081201(R)

[10] Hosen MM, Dhakal G, Wang B, Poudel N, Dimitri K, Kabir F, et al. Experimental observation of drumhead surface states in $SrAs_3$. Scientific Reports. 2020;**10**:2776

[11] Wang X, Ding G, Cheng Z, Surucu G, Wang X-L, Yang T. Novel topological nodal lines and exotic drum-head-like surface states in synthesized CsCl-type binary alloy TiOs. Journal of Advanced Research. 2020;**22**:137

[12] Li S, Yu Z-M, Liu Y, Guan S, Wang S-S, Zhang X, et al. Type-II nodal loops: Theory and material realization. Physical Review B. 2017;**96**:081106

[13] Bian G, Chang T-R, Zheng H, Velury S, Xu S-Y, Neupert T, et al. Drumhead surface states and topological nodal-line fermions in $TiTaSe_2$. Physical Review B. 2016;**93**:121113

[14] Weng H, Liang Y, Xu Q, Yu R, Fang Z, Dai X, et al. Topological node-line semimetal in three-dimensional graphene networks. Physical Review B. 2015;**92**:045108

[15] Neupane M, Belopolski I, Hosen MM, et al. Observation of topological nodal fermion semimetal phase in ZrSiS. Physical Review B. 2016; **93**:201104

[16] Topp A, Lippmann JM, Varykhalov A, et al. Non-symmorphic band degeneracy at the Fermi level in ZrSiTe. New Journal of Physics. 2016;**18**: 125014

[17] Zhao J, Yu R, Weng H, et al. Topological node-line semimetal in

compressed black phosphorus. Physical Review B. 2016;**94**:195104

[18] Xu G, Weng H, Wang Z, et al. Chern semimetal and the quantized anomalous hall effect in $HgCr_2Se_4$. Physical Review Letters. 2011;**107**:186806

[19] Nicol EJ, Carbotte JP. Optical conductivity of bilayer graphene with and without an asymmetry gap. Physical Review B. 2008;**77**:155409

[20] Fang C, Weng H, Dai X, Fang Z. Topological nodal line semimetals. Chinese Physics B. 2016;**25**:117106

[21] Zhou W, Rudenko AN, Yuan S. Effect of mechanical strain on the optical properties of nodal-line semimetal ZrSiS. Advanced Electronic Materials. 2020; **6**(1):1900860

[22] Carbotte JP. Optical response of a line node semimetal. Journal of Physics. Condensed Matter. 2017;**29**:045301

[23] Mukherjee SP, Carbotte JP. Transport and optics at the node in a nodal loop semimetal. Physical Review B. 2017;**95**:214203

[24] Barati S, Abedinpour SH. Optical conductivity of three and two dimensional topological nodal line semimetals. Physical Review B. 2017;**96**:155150

[25] Habe T, Koshino M. Dynamical conductivity in the topological nodal-line semimetal ZrSiS. Physical Review B. 2018;**98**:125201

[26] Wang C, Wen-Hui X, Zhu C-Y, Chen J-N, Zhou Y-L, Deng M-X, et al. Anomalous hall optical conductivity in tilted topological nodal-line semimetals. Physical Review B. 2021;**103**:165104

[27] Chan Y-H, Ching-Kai Chiu MY, Chou and Andreas P. Schnyder. Ca_3P_2 and other topological semimetals with line nodes and drumhead surface states. Physical Review B. 2016;**93**:205132

[28] Nag T, Menon A, Basu B. Thermoelectric transport properties of Floquet multi-Weyl semimetals. Physical Review B. 2020;**102**:014307

[29] Tabert C, J. Electronic Phenomena in 2D Dirac-Like Systems: Silicene and Topological Insulator Surface States. Guelph, Canada: The University of Guelph; 2015

Chapter 2

Graphene Nanosheets as Novel Nanofillers in an Epoxy Matrix for Improved Mechanical Properties

Hadja Imane Beloufa, Mostapha Tarfaoui, Khalid Lafdi, Mohamed Daly and Amine Bendarma

Abstract

In this paper, we will be interested in bending tests on a polymer matrix reinforced with graphene nanofillers. The mechanical behaviour and the damage kinetics were determined. The samples were made using controlled dispersions of graphene nanosheets (GNP) in EPON 862 matrix. Various samples with different contents of GNP were made (0, 0.5, 1, 2.5, 5 and 10% by weight). Mechanical properties such as maximum stress, strain at break and Young's modulus were determined. After each test, the fracture surfaces were characterised using optical microscopy (OP) and scanning electron microscopy (SEM). Experimental results show that the fracture toughness of the GNP/epoxy-based nanocomposites decreases with an increasing percentage of nanofillers. The flexural strength of the samples with 10 wt% of graphene significantly decreased compared to neat epoxy. Based on stress-strain data and the fracture surface analysis, graphene nanosheets seem to impact the mechanical behaviour and the kinetics of the damage. The influences of the weight percentage of GNP on the EPON matrix properties and the performance of the nanocomposites are discussed. In addition, the evolution of bending performance and damage kinetics with graphene content was obtained and analysed.

Keywords: epoxy matrix, graphene nanosheets, bending test, global properties, graphene effect

1. Introduction

Polymer-based composites were introduced in the 1960s as new structural materials. It disperses stiff, durable, and high-strength fibres in a polymer matrix. As a result, we fabricated lightweight composites [1]. Graphite nanosheets, carbon nanotubes (CNTs), and functionalised graphene have been used as new additives due to their excellent mechanical strength. In particular, graphene, a single layer of aromatic carbon, has one of the strongest among all materials and outstanding thermal and electrical properties [2–5]. The latest research on the impact of graphene sheets as a filler in polymers suggests that the addition of graphene can significantly improve the resulting composite materials' mechanical, thermal, and electrical properties.

 IntechOpen

The graphene filler can enhance the polymer's stiffness, strength, and toughness and increase its thermal and electrical conductivity [6]. Furthermore, integrating graphene into a polymer matrix can produce a new type of mechanical material by combining graphene's exceptional mechanical properties with the polymer's structural properties. Kuilla wrote an excellent state-of-the-art review article on "Recent advances in graphene-based polymer composites" [7].

Moreover, the dispersion of graphene in the polymer matrix is a critical factor in determining the properties of the resulting composite material. Researchers are exploring various methods for achieving a high degree of dispersion, such as graphene's chemical functionalization and surfactants' use.

The mechanical performance of composites with continuous fibre in a polymer matrix is well documented [8–10]. However, these composites have drawbacks that are essentially related to the intrinsic properties of the matrix and which subsequently limit their field of application [11, 12]. Recently, polymer-based nanocomposites have generated enormous interest because of their ability to improve the mechanical performance of epoxy resin [11, 13, 14]. Nanocomposites containing Nano additives like carbon black (CB), carbon nanotubes (CNTs) and graphene nanosheets (GNPs) offer excellent mechanical properties [15–17]. Many kinds of fillers, such as metal particles [18, 19], carbon nanofillers (CNF) [20, 21], carbon nanotubes [22, 23] and graphene [24–26], have been used. Graphene offers much better mechanical performance than other nanofillers, resulting in exceptional mechanical strength, significant elongation and large surface area [27, 28].

Mohan et al. [29] present a study of the impact of carbon allotropes on a self-powered triboelectric humidity sensor based on a flexible polydimethylsiloxane (PDMS) film. The authors used different carbon allotropes such as carbon nanotubes, graphene, and reduced graphene oxide as additives to the PDMS film. The study found that adding carbon allotropes to the PDMS film increased the humidity sensitivity of the sensor and improved its response time. The authors also observed an increase in the film's dielectric constant with the addition of carbon allotropes, which is attributed to the adsorption of moisture and the formation of an electric double layer. The results suggest that adding carbon allotropes to PDMS films can create a self-powered triboelectric humidity sensor with improved performance.

Ai et al. [30] explore the potential of using graphene/electrospun carbon nanofiber sponge composites induced by magnetic particles as a multi-functional pressure sensor. The composites were fabricated via electrospinning and then exposed to a magnetic field. The results showed that the resulting composite had a high sensitivity, good repeatability and durability, and a low hysteresis effect. Furthermore, the composite demonstrated excellent performance in terms of pressure sensing, temperature sensing, and humidity sensing. This demonstrates the potential of graphene/electrospun carbon nanofiber sponge composites for use in pressure-sensing applications.

Topal et al. [31] present a novel method to improve the performance of repaired secondary load-bearing aircraft composite structures. This technique employs scarf-bonded patches as part of a multiscale nano-integration strategy. This method preserves the structural integrity and durability of restored composite structures while minimising the loss of mechanical strength. This technique employs a nanomaterial-based adhesive that permanently attaches the scarf-bonded patches to the composite framework. The results of this study show that the performance of the repaired composite structures can be improved significantly by using this approach.

Wang et al. [32] have developed and studied different graphene nanofillers such as reduced graphene oxide, graphene nanosheets (GNP) and three-dimensional graphene. They demonstrated that epoxy reinforced with nanofillers has exceptional mechanical properties. In another work, an epoxy adhesive's mechanical properties and toughness were improved with graphene nano additives [33].

The epoxy matrix material is a composite's most commonly used primary phase because of lesser shrinkage, outstanding adhesion, and better solvent resistance. Some applications of epoxy-based nanocomposites include aerospace, automotive, marine, sporting goods, industrial tooling, and other general consumer products. To the best of the authors' knowledge, the effects of adding GNPs on the flexural properties of the epoxy matrix have not been reported in the literature. Herein, the epoxy matrix has been reinforced by graphene nanosheets (GNP) with different weight percentages. The effects of adding (GNPs) with different concentrations of GNPs (0, 0.5, 1, 2.5, 5 and 10% by weight) on the flexural properties of the polymer matrix were investigated. Enhanced GNPs-epoxy interface bonding is achieved using the three-roll mill. Homogeneous dispersion is further facilitated through sonication. The evolution of the flexural strength with the graphene content was obtained and analysed. The effect of nanofillers on the kinetics of damage is also reported.

2. Materials and methods

These samples include EPON Resin 862 epoxy. The graphite used in these studies was a natural flake with a 500 m diameter. First, exfoliated graphite (ExG) was manufactured using a mixture of nitric, sulphuric, and natural graphite. Intercalation with graphene sheets appears to produce an intercalated graphite product after 24 h of reaction. The mixture was subsequently filtered, rinsed with water, and dried at low temperatures in an oven. The intercalated graphite compound was then subjected to a hurried heat treatment at 900°C and a rapid extension.

Figure 1 displays the morphological characterisation of ExG nanofiller by SEM at three magnifications: 1 mm, 100 μm, and 20 μm. ExG nano-filler has a cellular structure associated with considerable expansion, as seen in SEM images. Before exfoliation, graphite exists as flakes with the graphite c-axis perpendicular to the plane of the flake. Due to the substantial expansion along the c-axis, the exfoliated flake becomes longer in the direction corresponding to the flake's c-axis prior to exfoliation. **Figure 1a** depicts the ExG generated from a graphite flake under low magnification (1 mm) and high magnification (100 μm). **Figure 1b** depicts the ExG under high magnification (100 μm). **Figure 1c** shows graphene sheets with thicknesses varying from 1 nm to around 16 nm and widths ranging from sub-micrometre to hundreds of m at 20 m magnification. **Figure 1** depicts that the gap between graphitic stacks is roughly 5–10 nm.

To fabricate samples, a traditional method of dry mixing and adding a different mass percentage of graphene (0.5, 1, 2.5, 5 and 10 wt%) to the epoxy resin was used. The reference polymer and the GNP/epoxy nanocomposites were treated under the same conditions to ensure the samples' homogeneity. This study used the exfoliation and reduction method to produce graphene, **Figure 1d**. The synthesis of the product requires the use of Epon 862 resin and the hardener Epikure W. This epoxy matrix was selected for its characteristics at ambient temperatures, such as low viscosity and ease of handling. In addition, the chosen hardener offers good chemical resistance at high temperatures and excellent performance. **Figure 2** shows the molecular formulae of both compounds. One of the big problems related to the synthesis of graphene and

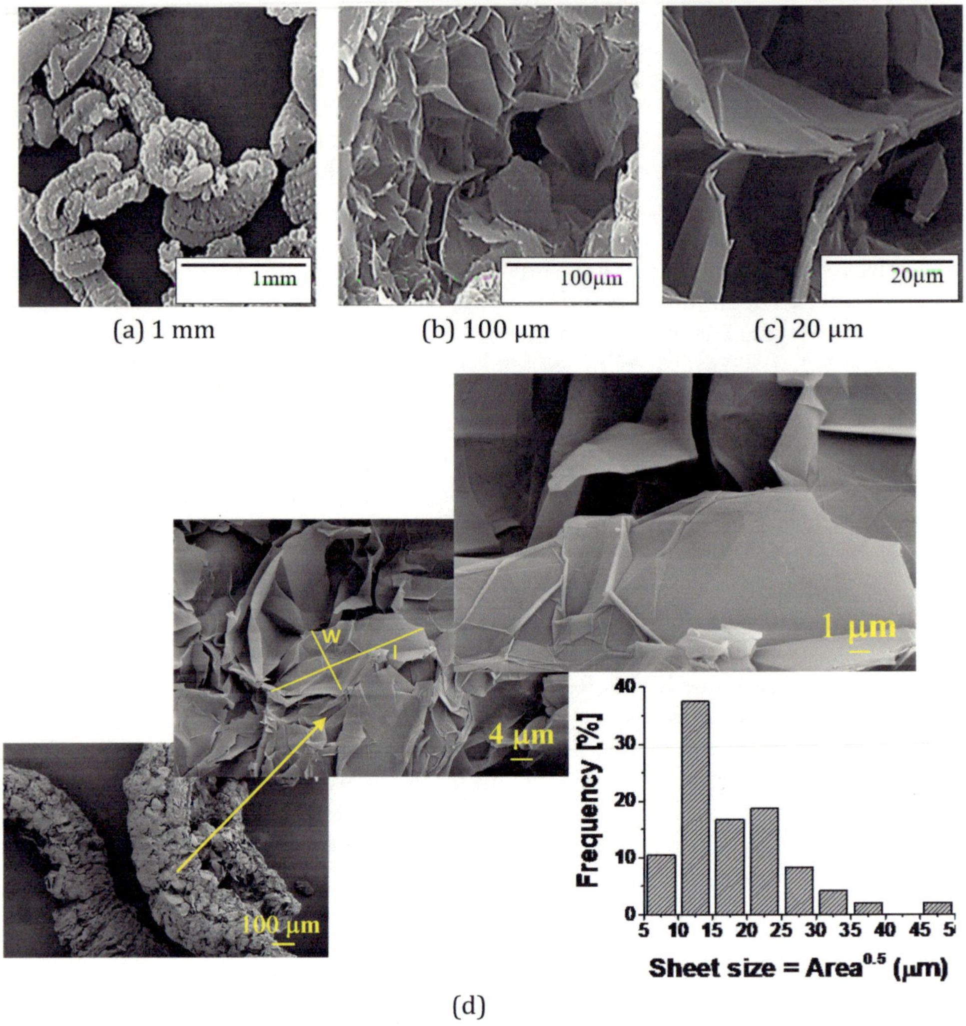

Figure 1.
Size distribution and SEM image of the graphene nanosheets.

its dispersion in a polymer is the phenomenon of agglomeration. To overcome this problem and a dry mixing process was used.

Six concentrations of GNP/epoxy (0, 0.5, 1, 2.5, 5 and 10 wt%) were used. The control sample consists of pure resin. It has been noted that for concentrations greater than 10% by weight, the mixing process becomes ineffective and difficult to obtain a well-dispersed filler.

GNPs were mixed with the resin/hardener solution for the desired concentration using a three-roll mixer. The same procedure was adopted for all percentages except for 10 wt% because the mixture's viscosity increases as the number of nanofillers increases. Therefore, several steps are required to prepare the specimen at 10 wt%. First, a small percentage of graphene is mixed with the resin using the three-roll mill until the mixture's viscosity is reduced, and then we add more GNP. This step was repeated several times until all the graphenes were mixed uniformly in the

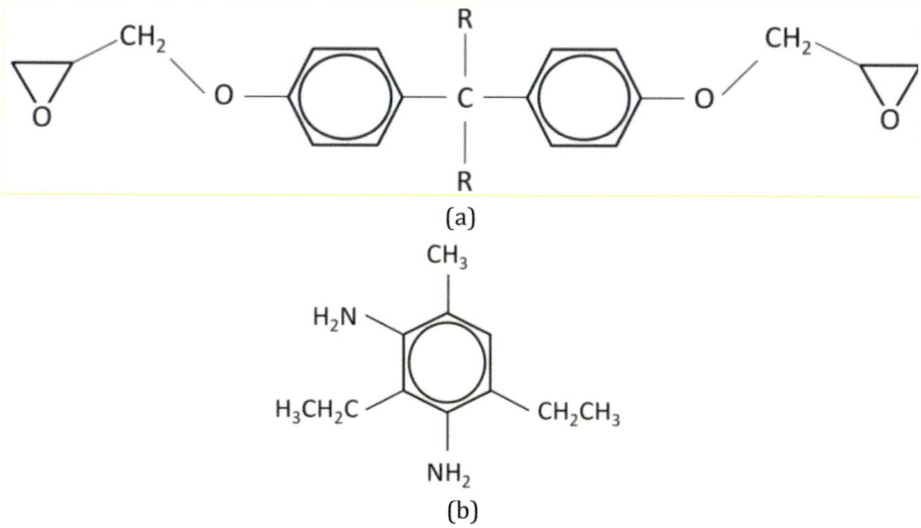

Figure 2.
Molecular formulation of EPON Resin 862. (a) Epon 862 Resin (b) Epikure W Hardener.

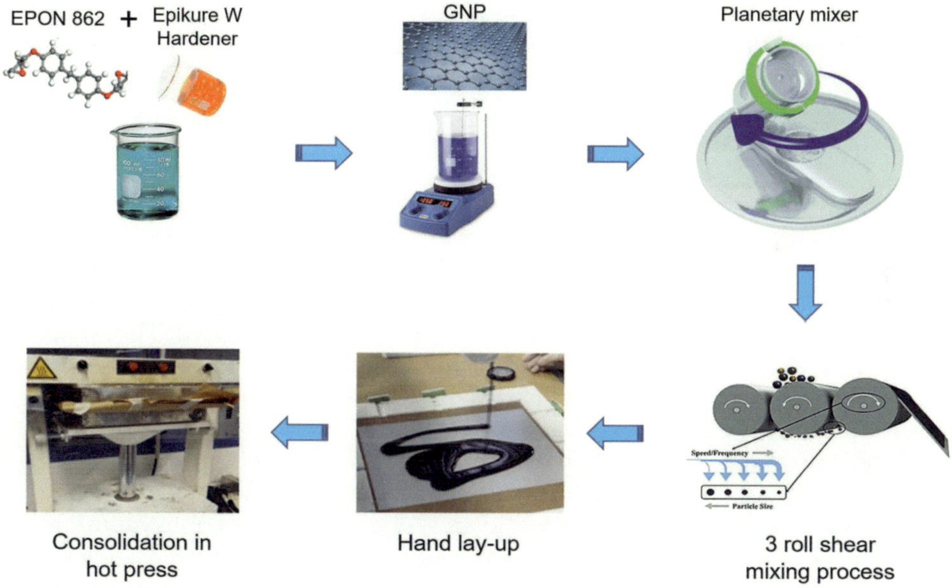

Figure 3.
Schematic diagram of GNP/epoxy nanocomposite production fabrication.

resin, **Figure 3**. Based on previous studies, the three-roll mill is the most appropriate apparatus to obtain a uniform GNP/epoxy mixture [28].

The speeds of the three cylinders 1, 2, and 3 are 21, 64, and 200 rpm, respectively. The space between the rolls was kept small to produce significant shear stress. Here, the distance between the rolls was fixed to 25 μm. It was decided to pass the mixture several times in the mill for all formulations. As graphene content increases, so does the speed of the feed roll, **Table 1**. Since there is an increase in the

Formula	GNP (%)	Velocity (rpm)
1	0	333
2	0.5–1	400
3	2.5	480
4	5	550
5	10	600

Table 1.
Velocity of the feed roller.

	GNP (%)	Epon 862
Young Modulus (GPa)	1030	2.72
Poisson coefficient	0.19	0.3

Table 2.
Properties of epoxy and graphene [29–33].

temperature in the system, which is generated by the shear stress, the temperature of the rolls was maintained at a temperature of 21°C using a cooling system (**Figure 3**).

Table 2 shows the mechanical properties of graphene nanosheets and epoxy matrix.

3. Results and discussion

This study made composite specimens consisting of an Epoxy matrix reinforced with GNP. Six specimens with 0, 0.5, 1, 2.5, 5 and 10% of GNP were manufactured and experimentally tested under Bending (ASTM D2344). The mechanical bending performances of the epoxy matrix reinforced by graphene nanofillers were evaluated. Specimens with L = 50 mm, W = 15 mm, and T = 5 mm has been used. The schematic representation of the dimensions of the specimens studied, and the bending device is given in **Figure 4**. The specimens are placed on the two supports 10 mm in diameter and are separated by a fixed distance.

A bending test was used to study the mechanical behaviour of neat epoxy and GNP/epoxy nanocomposites. In addition, a bending test was used to assess the effect of nano-reinforcement on the mechanical characteristics of epoxy reinforced with graphene nanosheets. The three-point bending test was conducted using the universal testing machine (Instron type 5585H). The load frame was equipped with Instron's ± 50 kN static cell. All measurements were taken at room temperature, and all tests were performed at a speed of 0.5 mm/min. The mechanical properties of manufactured specimens were evaluated for each GNPs fraction using the bending test's force-displacement and stress-strain curves. For reproducibility of results, all specimens were subjected to a set of bending tests (3 tests).

3.1 GNPs distribution in an epoxy matrix

In this work, the distribution of GNP in the epoxy matrix was analysed using SEM and optical analysis. **Figure 5a** shows optical microscope images of graphene-filled

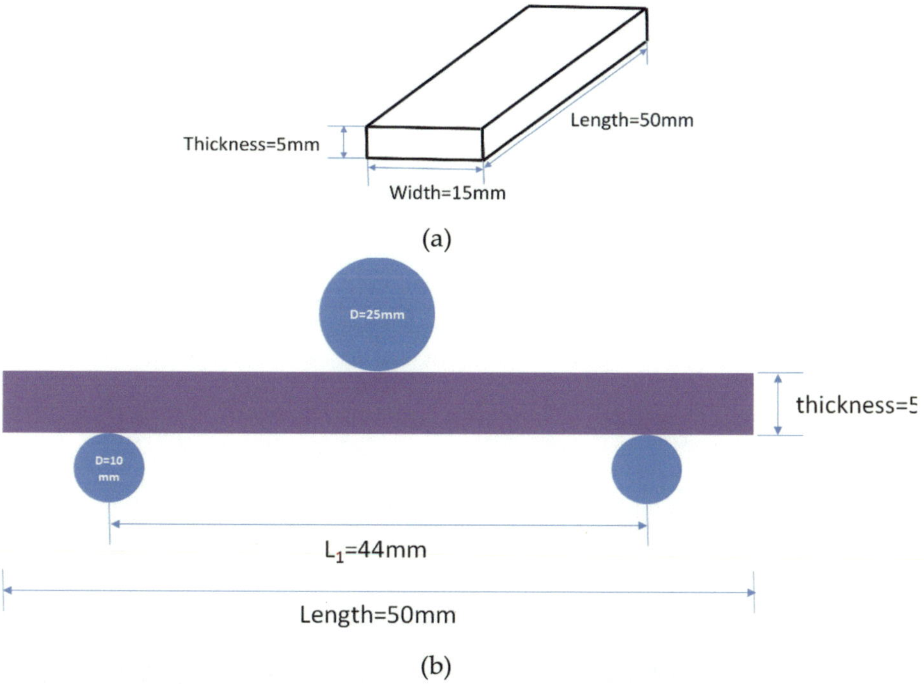

Figure 4.
Schematic presentation of test specimen dimensions and bending test. (a) Dimension of specimens (b) Bending test.

epoxy. Arrows indicate the graphene sheets. In all cases, one can see that the graphene nanosheets are well dispersed all over the surface. **Figure 5** also shows that the dispersion of GNP in epoxy is reasonably uniform. **Figure 5** gives an overview of the GNP/epoxy nanocomposites identifying the graphene sheets, while **Figure 6** shows a zoom with a detailed view. **Figure 5b** illustrates the surfaces of GNP/epoxy nanocomposite, showing the graphene morphology and their dispersions into a polymer matrix. The graphene platelets were exfoliated, exhibiting a wide range of interatomic interlayer spacing. They are well dispersed with no evidence of aggregates. This morphology has an essential role in the mechanical performance and load transfer from the epoxy matrix to graphene sheets, as reported by*** Shen et al. [34].

3.2 Bending properties of GNP/epoxy nanocomposites: effect of GNPs

GNPs were randomly distributed in the Epoxy matrix. The mechanical properties of the GNP/epoxy samples were determined for each mass fraction using a flexural test. For each GNP/epoxy configuration, five tests per configuration were carried out to ensure the reproducibility of the tests. **Figure 6** illustrates the reproducibility of obtained results. The elastic moduli, peak stress, strain at break, and stiffness were evaluated for each test.

Figure 7 summarises the results of the tests carried out on the specimens with different percentages of graphene: 0.5, 1, 2.5, 5, and 10 wt%. Again, the effect of nanofillers on the overall behaviour of the material is apparent and noticeable.

As shown in **Figure 8**, the resin with added nanofillers has a lower resistance than the neat epoxy. The variation of the bending properties (Peak stress, Strain at

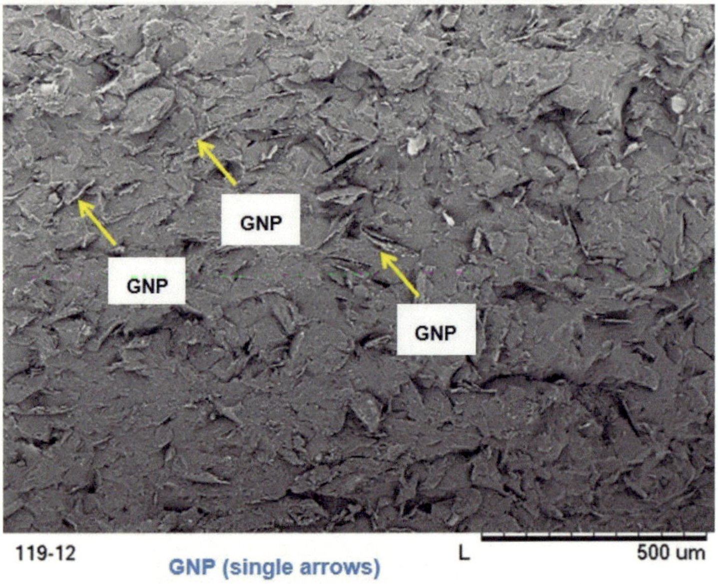

(a)

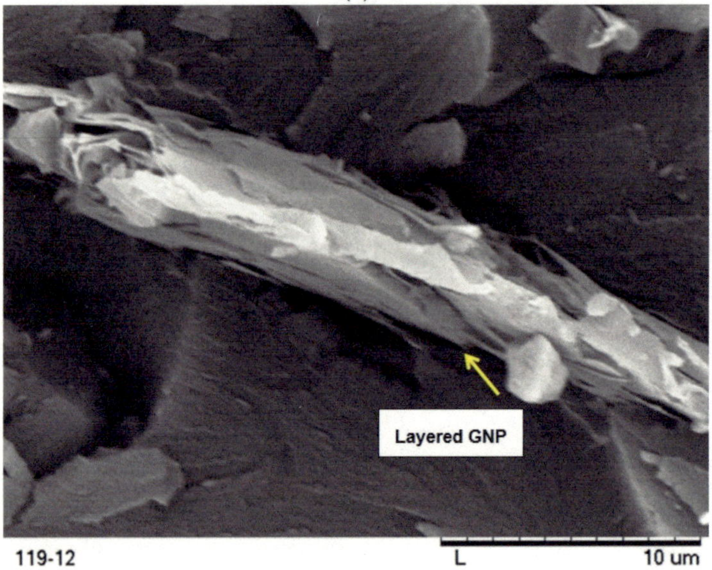

(b)

Figure 5.
Overview of GNP/epoxy nanocomposite surfaces. (a) Distribution of GNP in epoxy matrix and (b) zoom of GNP/epoxy nanocomposite and graphene morphology.

break, Stiffness) of GNP/epoxy specimens with graphene content is shown in **Figures 8** and **9**. As expected, the stiffness of the specimens increased with the increase of the mass fraction of the nanofillers. However, the peak stress and the strain at break (or strain at maximum stress) dropped when adding graphene nanofillers. The improvement can be attributed to GNP's high flexural modulus. The nanocomposite reinforced with 1% GNP demonstrated a more significant increase

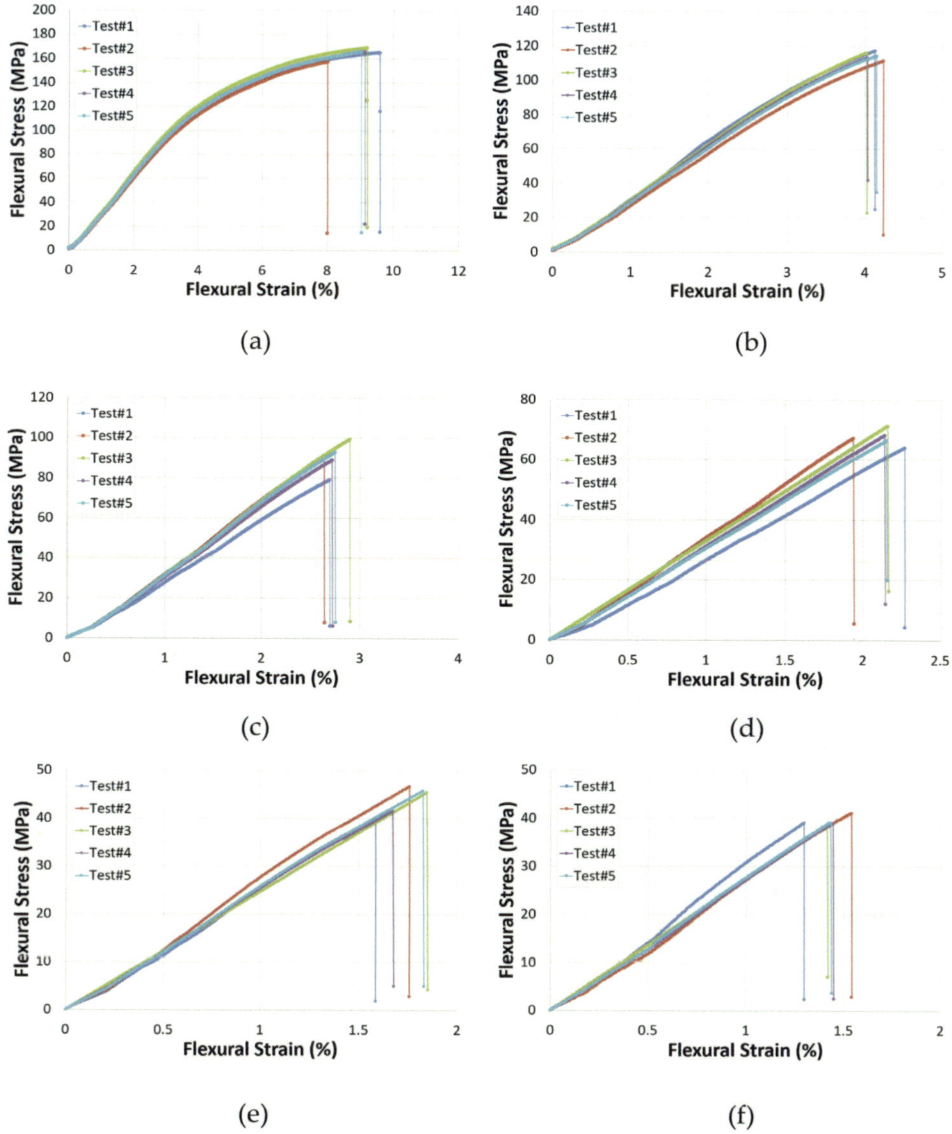

Figure 6.
Stress-strain curves for a different formulation of GNP/epoxy composites. (a) GNP, 0 wt%, (b) GNP, 0.5 wt%, (c) GNP, 1 wt%, (d) GNP, 2.5 wt%, (e) GNP, 5 wt% and (f) GNP, 10 wt%

in elastic modulus, as seen in **Table 3**. The GNP/epoxy mixture with 5% GNP has a slightly lower flexural modulus than pure epoxy. For 10% GNPs, the addition of graphene nanosheets has almost no effect on the stiffness of the base material.

The bending strength of the GNP/epoxy nanocomposites decreased peak stress and strain at break, **Figures 6** and **7**. This degradation has already been noted in some research [35–37], which has explained the poor dispersion of graphene. The formation of aggregates can cause a stress concentration. The interfacial strength controls the ultimate mechanical properties. Thus the weak interfacial adhesion between GNPs and the epoxy matrix and the aggregation of nanosheets explain the strength reduction. As can be seen in **Tables 4** and **5**, the resistance of nanocomposites decreased

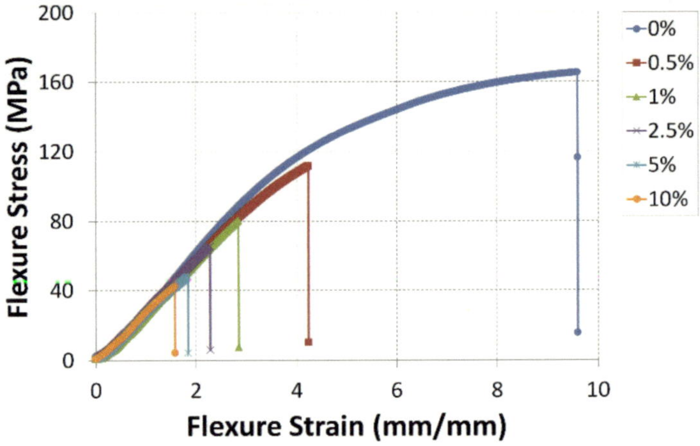

Figure 7.
Bending behaviour of different formulations of GNP/epoxy composites.

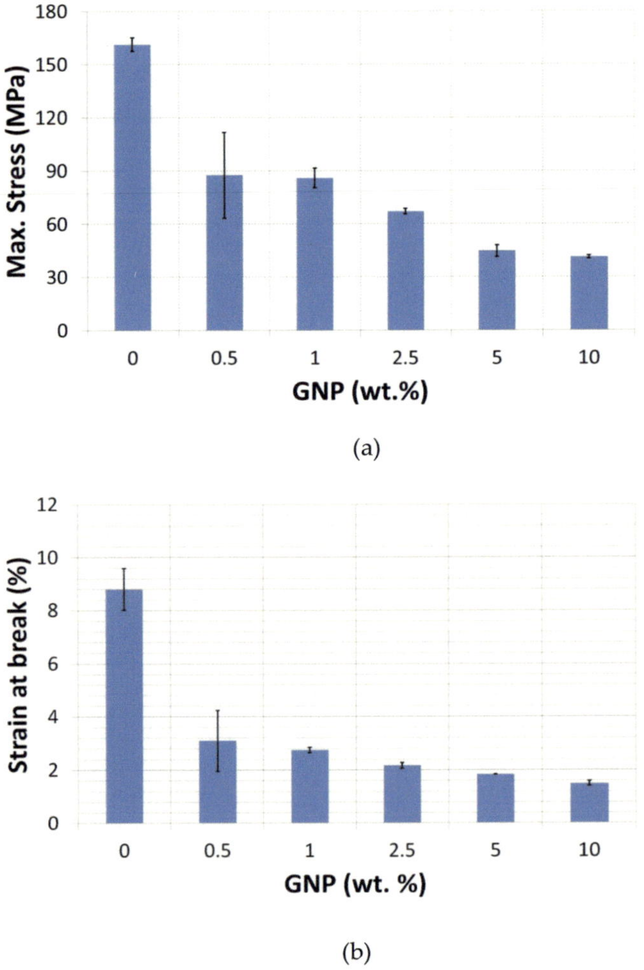

Figure 8.
Bending properties of GNP/epoxy composites. (a) Peak stress vs. wt.% GNP and (b) strain at break vs. wt% GNP.

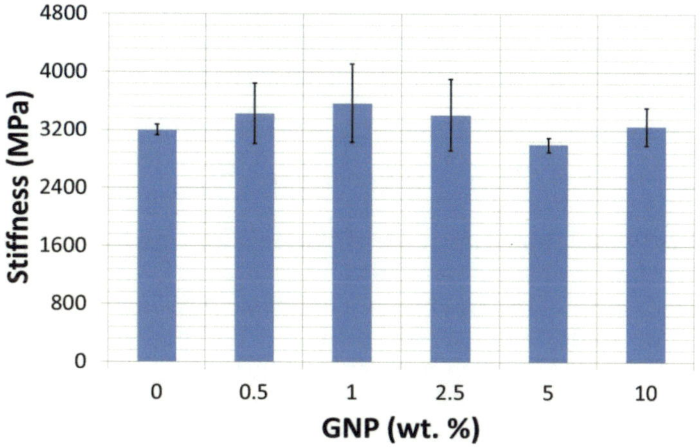

Figure 9.
Stiffness of GNP/epoxy composites.

GNP (wt%)	0	0.5	1	2.5	5	10
Average (MPa)	3203.12	3425.65	3569.57	3410.22	2998.21	3250.36
Performance (%)	—	+6.95	+11.44	+6.47	−6.40	+1.47

Table 3.
Flexural modulus variation vs. wt% GNPs.

GNP (wt%)	0	0.5	1	2.5	5	10
Average (MPa)	161.28	87.54	85.93	67.12	44.86	41.43
Performance (%)	0	−45.72	−46.72	−58.38	−72.19	−74.31

Table 4.
Peak stress variation vs. wt% GNPs.

GNP (wt%)	0	0.5	1	2.5	5	10
Average (MPa)	8.802	3.0955	2.7445	2.169	1.839	1.4965
Performance (%)	0	−64.83	−68.82	−75.36	−79.11	−83

Table 5.
Strain at break variation vs. wt% GNPs.

with the increase in the GNP contents. As expected, the presence of GNPs led to more fragile materials because the peak load and the stress at break are conditioned by the interactions between the phases and the chemical bonds between the graphene and the epoxy matrix, which have been confirmed by different studies [38].

Tables 4 and **5** show the average peak stress and strain at break versus GNP mass fraction. Adding GNPs significantly affects the peak stress and strain at failure, which decreases with the addition of GNPs. The peak stress decreases by 45.72, 46.72, 58.38, 72.19 and 74.31% for 0.5, 1, 2.5, 5 and 10 wt%, respectively. As presented in **Table 1**, the elastic part of the curve was slightly higher for samples containing 0.5, 1, and 2.5 wt% GNP than the neat epoxy. Considering the standard deviation, the stiffness of all specimens is nearly consistent, **Figure 8**. However, the material's stiffness decreased when the contents of GNP exceeded 2.5%. It can be concluded that, in the case of bending tests, the addition of GNPs causes a decrease in mechanical performance.

3.3 Bending resistance of GNP/epoxy nanocomposites

Figure 8 and **Tables 2** and **3** show the evolution of the mechanical properties for the different specimens with different graphene contents. The five nanocomposites (GNP/epoxy) have maximum stress and a maximum strain lower than the neat epoxy. Indeed, under three-point bending tests, the higher the graphene content, the more it weakens the material (the bending strength continues to decrease).

After each test, the fracture surface of specimens was examined by an optical microscope (Kayence) and scanning electron microscope (MEB). **Figure 10** visualises the different damage mechanisms in the Kayence and MEB micrographs of the fractured specimen (central region), which are activated during the bending test of the manufactured composite. From these results, it is clear that, under bending stress, these materials (matrix doped with GNP) have an unfortunate tendency to be damaged rapidly compared to neat epoxy (their performance decreases); this is the result of changes in their structure. By examining these various evolutions, we will be able to describe and classify precisely the various types of damage to materials. The embrittlement is the outcome of internal material changes that are not accompanied by the formation of new surfaces but lower the material's ductility. The first explanation for the decrease in flexural strength as the graphene content increases is the aggregation of GNP in the epoxy matrix, **Figure 11**. The second reason is that nanofillers tend to form a barrier for the deformation of the specimens, which leads to a faster failure with too low levels of load and deformation.

Microscopic analyses of the GNP/epoxy samples were carried out to understand the relationship between their morphology and mechanical behaviour. The analysis of

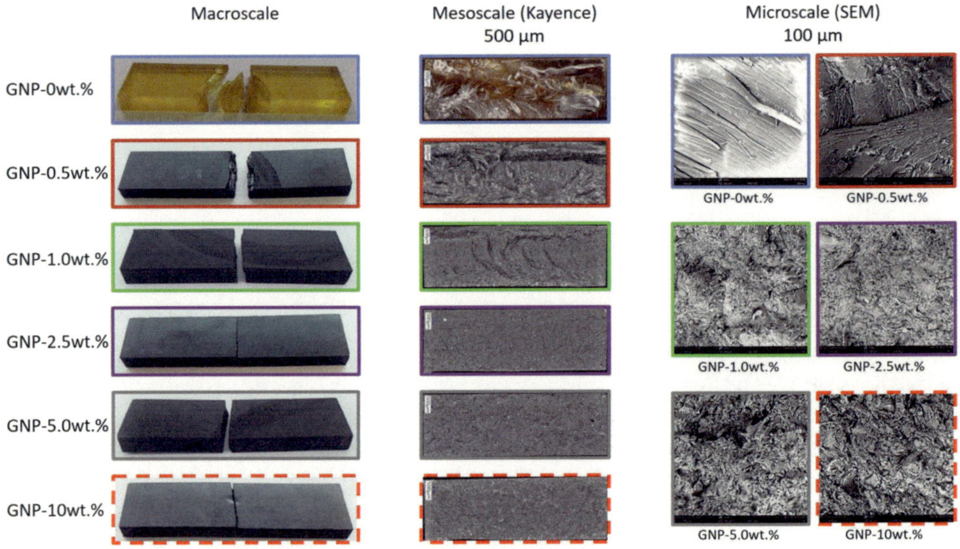

Figure 10.
Fracture facies, GNP/epoxy with different percentage of GNP at different scales.

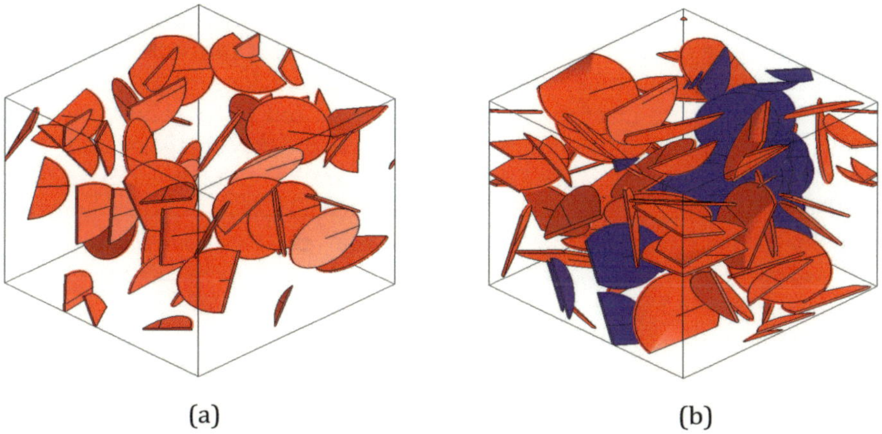

Figure 11.
Simulation of GNP distribution in the epoxy matrix. (a) 1 wt% GNP and (b) 5 wt% GNP—with agglomeration.

the stress-strain curves, as well as the fracture surfaces, allows us to state the following points:

- Dispersion of nanofillers in a reference material is a recurrent problem of nanocomposites, but it is more visible for a high percentage. This dispersion can cause the formation of agglomeration (**Figure 11**) and poor adhesion, which will cause the fall of the performances of the nanocomposites. So, we can say that the more the nanofiller percentage increases, the more this problem is present.

- The adhesion of the GNPs to the matrix is not perfect, **Figure 10**. As previously stated, graphene nanofillers act as a barrier to crack propagation and limit the deflection of the specimens. The kinetics of damage, unfortunately, does not favour the hardening of these nanocomposites because of the weakness of the GNP/epoxy interface.

- The bending test of the GNP/epoxy results in various damage mechanisms responsible for the suppression of crack initiation and propagation in the composite. As the graphene content in the epoxy matrix increases, the propagating crack begins to interact with the rigid GNP; it must tilt or twist at greater angles to pass the GNP due to the high aspect ratio of the particles. As a result, the fracture surface becomes rougher with more peaks and valleys.

- From the bending test, it is clear that the dispersion and doping of the epoxy matrix with different mass ratios of graphene nanoparticles has an effect not only on the load/stress at break and the bending modulus but also on the mode of failure. Indeed, the behaviour of the GNP-doped epoxy makes the material more brittle with a plastic zone that reduces with increasing GNP mass percentage. For 10% of nanofillers, the behaviour is purely brittle; failure occurs at the end of the elastic domain.

- The results show that the increase in the GNP mass fraction degrades the properties of the epoxy; they did not improve the epoxy performance. In **Figure 10**, the crack guidance mechanism can also be visualised. In crack guidance, cracks propagate along the GNPs, and lower fracture energy is consumed for the reinforced epoxy; this is because the GNPs hinder crack propagation in the load plane due to their unique wrinkling characteristic and strongly limit the plastic zone. In GNP-epoxy nanocomposites, the higher the percentage of GNP, the more localised the crack propagation path is (parallel to the direction of the applied load). This phenomenon is attributed to the robust bonding interfaces between the GNPs and the matrix grains. In contrast, extensive plastic deformation is observed for pure epoxy, leading to a bifurcation or branching of the crack, as shown in **Figure 10**.

- Two situations are to be considered (a) for low plastic deformations of the matrix, the sliding is still very heterogeneous, and the Zener mechanism of dislocation stacking can be considered on the obstacle constituted by the GNP inclusion (b) for more significant plastic deformations, the field of these deformations becomes much more homogeneous, in which case the mechanics developed by Eshelby provides answers.

- For this type of test and material, we can consider that the stress concentration in the inclusion is an increasing function of the strain-hardening slope and depends on the shape factor of the inclusion. This factor takes exceptionally high values for needle-shaped inclusions parallel to the maximum principal stress and disc-shaped inclusions stressed in their plane. We understand that such inclusions are those that cleave or whose interface breaks first. Hence, they are particularly harmful from the point of view of the damage; this is the case, for example, of GNP inclusions in the epoxy matrix.

- The distribution of GNP (volume fraction of inclusions) is far from constant within a material. Instead, we often find clusters of high volume fraction (agglomeration effect). The damage develops primarily in these zones. While this more or less random distribution of inclusions has a negligible influence on the elastic behaviour of the material, it plays an essential role in the damage. Once again, disorder favours it.

4. Conclusions

In this work, we studied the effect of graphene nanofillers (GNP) on the bending performance of an epoxy matrix. Six weight percentages of graphene nanosheets (0, 0.5, 1, 2.5, 5 and 10 wt%) were prepared and tested. At this level, the first conclusion is that flexural modulus increased with increasing GNP, reaching a maximum value of 1 wt% (an increase of 12%). However, the flexural strength of the nanocomposites decreased as the percentage of graphene increased, which can be explained by the formation of graphene aggregates and could also be the result of the interfacial bonding between the epoxy matrix and the GNP nanosheets. Toughness (peak stress and strain at break) drops with increasing graphene content for all nanocomposite formulations. Nanocomposites with 10 wt% of GNP are the least resistant to flexural stress. However, GNP/epoxy resulted in a higher increase in the flexural modulus with a maximum of 1 wt% of nanofillers. GNP/epoxy nanocomposites have a higher modulus than neat epoxy due to fewer crosslinked networks. The shear strength for nanocomposites at 0.5% GNP is greater than for 10% GNP, which is a consequence of agglomerates as the GNP content increases. The chemical bonds formed between the epoxy matrix and GNP had a major effect on the mechanical behaviour and the strength of specimens.

To our knowledge, and after analysis of the bibliographic references, there is no work dedicated to quantifying the performances of a matrix doped by nanofillers under 3-point bending type solicitations. Nevertheless, this work constitutes an exciting database for scientists and industrialists and complements our work on nanocomposites and understanding their mechanical responses and damage kinetics under static and dynamic loads [39–48].

Conflict of interest

The authors declare no conflict of interest.

Author details

Hadja Imane Beloufa[1,2], Mostapha Tarfaoui[1,3]*, Khalid Lafdi[4,5], Mohamed Daly[1,6] and Amine Bendarma[7]

1 ENSTA Bretagne, IRDL, UMR CNRS 6027, Brest, France

2 FST, Ibn Badis University Mostaganem, Mostaganem, Algeria

3 Green Energy Park (IRESEN/UM6P), Benguerir Morocco

4 University of Dayton, Dayton, Ohio, USA

5 Northumbria University, Newcastle, United Kingdom

6 ENISo University of Sousse, Sousse, Tunisia

7 Polytechnic School of Agadir, Morocco

*Address all correspondence to: mostapha.tarfaoui@ensta-bretagne.fr

References

[1] Rajak DK, Pagar DD, Menezes PL, Linul E. Fiber-reinforced polymer composites: Manufacturing, properties, and applications. Polymers. 2019;**11**(10): 1667

[2] Jesuarockiam N, Jawaid M, Zainudin ES, Hameed T, Sultan M, Yahaya R. Enhanced thermal and dynamic mechanical properties of synthetic/natural hybrid composites with graphene nanoplateletes. Polymers. 2019;**11**(7):1085

[3] Mishra RK, Goel S, Nezhad YH. Computational prediction of electrical and thermal properties of graphene and BaTiO3 reinforced epoxy nanocomposites. Biomaterials and Polymers Horizon. 2022;**1**(1):1-14

[4] Mousavi SR, Estaji S, Kiaei H, Mansourian-Tabaei M, Nouranian S, Jafari SH, et al. A review of electrical and thermal conductivities of epoxy resin systems reinforced with carbon nanotubes and graphene-based nanoparticles. Polymer Testing. 2022;**2022**:107645

[5] Lopez-Polin G, Gomez-Navarro C, Gomez-Herrero J. The effect of rippling on the mechanical properties of graphene. Nano Materials Science. 2022;**4**(1):18-26

[6] El Moumen A, Tarfaoui M, Hassoon O, Ben Yahia H, Lafdi K. Experimental study and numerical modelling of low velocity impact on laminated composite reinforced with thin film made of carbon nanotubes. Applied Composite Materials. 2017;**25**(2):309-320

[7] Kuilla T, Bhadra S, Yao D, Kim NH, Bose S, Lee JH. Recent advances in graphene based polymer composites. Progress in Polymer Science. 2010;**35**(11):1350-1375

[8] Hassoon OH, Tarfaoui M, El Moumen A. Progressive damage modeling in laminate composites under slamming impact water for naval applications. Composite Structures. 2017;**167**:178-190

[9] Tarfaoui M, Khadimallah H, Imad A, Pradillon JY. Design and finite element modal analysis of 48m composite wind turbine blade. Applied Mechanics and Materials. 2012;**146**:170-184

[10] Tarfaoui M, El Moumen A, Lafdi K. Progressive damage modeling in carbon fibers/carbon nanotubes reinforced polymer composites. Composites Part B: Engineering. 2017;**112**:185-195

[11] El Moumen A, Tarfaoui M, Lafdi K. Mechanical characterisation of carbon nanotubes based polymer composites using indentation tests. Composites Part B: Engineering. 2017;**114**:1-7

[12] Tarfaoui M, El Moumen A, Lafdi K. Mechanical properties of carbon nanotubes based polymer composites. Composites Part B: Engineering. 2016;**103**:113-121

[13] El Moumen A, Tarfaoui M, Lafdi K, BenYahia H. Dynamic properties of carbon nanotubes reinforced carbon fibers/epoxy textile composites under low velocity impact. Composites Part B: Engineering. 2017;**125**:1-8

[14] Ben Yahia H, Tarfaoui M, Datsyuk V, El Moumen A, Trotsenko S, Reich S. Dynamic properties of hybrid composite structures based multiwalled carbon nanotubes. Composite Science and Technology. 2017;**148**:70-79

[15] Tarfaoui M, Lafdi K, Beloufa I, Daloia D, Muhsan A. Effect of graphene nano-additives on the local

mechanical behavior of derived polymer nanocomposites. Polymers. 2018;**10**(6):667

[16] Tarfaoui M, El Moumen A, Boehle M, Shah O, Lafdi K. Self-heating and deicing epoxy/glass fibre-carbon nanotubes buckypaper composite. Journal of Materials Science. 2018;**54**(2):1351-1362

[17] El Moumen A, Tarfaoui M, Hassoon OH, Lafdi K, Benyahia H. Experimental study and numerical modelling of low velocity impact on laminated composite reinforced with thin film made of carbon nanotubes. Applied Composite Materials. 2018;**25**(2):309-320

[18] Ilyas RA, Sapuan SM, Asyraf MRM, Dayana DAZN, Amelia JJN, Rani MSA, et al. Polymer composites filled with metal derivatives: A review of flame retardants. Polymers. 2021;**13**(11):1701

[19] Tan NPB, Lee CH, Li P. Green synthesis of smart metal/polymer nanocomposite particles and their tuneable catalytic activities. Polymers. 2016;**8**(4):105

[20] Saba N, Safwan A, Sanyang ML, Mohammad F, Pervaiz M, Jawaid M. Thermal and dynamic mechanical properties of cellulose nanofibers reinforced epoxy composites. International Journal of Biological Macromolecules. 2017;**102**:822-828

[21] Ravindran AR, Ladani RB, Wu S, Kinloch AJ, Wang CH, Mouritz AP. Multiscale toughening of epoxy composites via electric field alignment of carbon nanofibres and short carbon fibres. Composites Science and Technology. 2018;**167**:115-125

[22] Mecklenburg M, Mizushima D, Ohtake N, Bauhofer W, Fiedler B, Schulte K. On the manufacturing and electrical and mechanical properties of ultra-high wt.% fraction aligned MWCNT and randomly oriented CNT epoxy composites. Carbon. 2015;**91**:275-290

[23] Liew KM, Lei ZX, Zhang LW. Mechanical analysis of functionally graded carbon nanotube reinforced composites: A review. Composite Structures. 2015;**120**:90-97

[24] Wang Z, Jia Z, Feng X, Zou Y. Graphene nanoplatelets/epoxy composites with excellent shear properties for construction adhesives. Composites Part B: Engineering. 2018;**152**:311-315

[25] Tang LC, Wan YJ, Yan D, Pei YB, Zhao L, Li YB. The effect of graphene dispersion on the mechanical properties of graphene/epoxy composites. Carbon. 2013;**60**:16-27

[26] Chandrasekaran S, Seidel C, Schulte K. Preparation and characterisation of graphite nanoplatelet (GNP)/epoxy nanocomposite: Mechanical, electrical and thermal properties. European Polymer Journal. 2013;**49**(12):3878-3888

[27] Miculescu M, Thakur VK, Miculescu F, Voicu SI. Graphene-based polymer nanocomposite membranes: A review. Polymers Advanced Technologies. 2016;**27**:844-859

[28] Anwar Z, Kausar A, Rafique I, Muhammad B. Advances in epoxy/graphene nanoplatelet composite with enhanced physical properties: A review. Polymer-Plastics Technology and Engineering. 2016;**55**(6):643-662

[29] Mohan V, Mariappan VK, Pazhamalai P, Krishnamoorthy K, Kim SJ. Unrevealing the impact of carbon allotropes in flexible polydimethylsiloxane film towards self-powered triboelectric humidity sensor. Carbon. 2023;**205**:328-335

[30] Ai J, Cheng SR, Miao YJ, Li P, Zhang HX. Graphene/electrospun carbon nanofiber sponge composites induced by magnetic particles for mutil-functional pressure sensor. Carbon. 2023;**205**:454-462

[31] Topal S, Al-Nadhari A, Yildirim C, Beylergil B, Kan C, Unal S, et al. Multiscale nano-integration in the scarf-bonded patches for enhancing the performance of the repaired secondary load-bearing aircraft composite structures. Carbon. 2023;**204**:112-125

[32] Wang Z, Shen X, Garakani MA, Lin X, Wu Y, Liu X. Graphene aerogel/epoxy composites with exceptional anisotropic structure and properties. ACS Applied Materials & Interfaces. 2015;7:5538-5549

[33] Quan D, Carolan D, Rouge C, Murphy N, Ivankovic A. Mechanical and fracture properties of epoxy adhesives modified with graphene nanoplatelets and rubber particles. International Journal of Adhesion and Adhesives. 2018;**81**:21-29

[34] Gojny FH, Wichmann MHG, Kopke U, Fiedler B, Schulte K. Carbon nanotube-reinforced epoxy-composites: Enhanced stiffness and fracture toughness at low nanotube content. Composites Science and Technology. 2004;**64**:2363-2371

[35] Chihi M, Tarfaoui M, Qureshi Y, Benyahia H, Bouraoui C. Graphene nanofillers as a player to improve the dynamic compressive response and failure behavior of carbon/epoxy composite. Nanotechnology. 2020;**31**:425709

[36] Ximenes FX. Durability of Fibre Reinforced Polymer (FRP) Composite Pipe; Faculty of Engineering. Porto, Portugal: University of Porto; 2017

[37] Koo JH, editor. Basics of polymer matrices and composites. In: Fundamentals, Properties, and Applications of Polymer Nanocomposites. Cambridge, UK: Cambridge University Press; 2016. pp. 109-129

[38] Rouway M, Nachtane M, Tarfaoui M, Chakhchaoui N, Omari LEH, Fraija F, et al. Mechanical properties of a biocomposite based on carbon nanotube and graphene nanoplatelet reinforced polymers: Analytical and numerical study. Journal of Composites Science. 2021;**5**(9):234

[39] Khammassi S, Tarfaoui M, Qureshi Y, Benyahia H. Mechanical properties of graphene nanoplatelets reinforced epikote 828 under dynamic compression. Mechanics of Materials. 2021;**158**:103873

[40] Shen X, Pei X, Fu S, Friedrich K. Significantly modified tribological performance of epoxy nanocomposites at very low graphene oxide content. Polymer. 2013;**54**:1234-1242

[41] Yasmin A, Daniel IM. Mechanical and thermal properties of graphite platelet/epoxy composites. Polymer. 2004;**45**(24):8211-8219

[42] Prolongo SG, Moriche R, Jiménez-Suarez A, Sanchez M, Urena A. Advantages and disadvantages of the addition of graphene nanoplatelets to epoxy resins. European Polymer Journal. 2004;**61**:206-214

[43] Prolongo MG, Salom C, Arribas C, Sanchez-Cabezudo M, Masegosa RM, Prolongo SG. Influence of graphene nanoplatelets on curing and mechanical properties of graphene/epoxy nanocomposites. Journal of Thermal Analysis and Calorimetry. 2016;**125**(2):629-636

[44] Prolongo MG, Gude MR, Ureña A. Improving the flexural and thermomechanical properties of amino-functionalised carbon nanotube/epoxy composites by using a pre-curing

treatment. Composites Science and Technology. 2011;**71**:765-771

[45] El Moumen A, Tarfaoui M, Nachtane M, Lafdi K. Carbon nanotubes as a player to improve mechanical shock wave absorption. Composites Part B: Engineering. 2019;**164**:67-71

[46] Khammassi S, Tarfaoui M, Lafdi K. Study of mechanical performance of polymer nanocomposites reinforced with exfoliated graphite of different mesh sizes using micro-indentation. Journal of Composite Materials. 2021;**2021**:0021

[47] Qureshi Y, Tarfaoui M, Lafdi KK, Lafdi K. Development of microscale flexible nylon/Ag strain sensor wire for real-time monitoring and damage detection in composite structures subjected to three-point bend test. Composites Science and Technology. 2019;**181**:107693

[48] Khammassi S, Tarfaoui M. Micromechanical characterisation of carbon black reinforced adhesive nanocomposite using micro indentation. Materials Today: Proceedings. 2021;**2021**

Chapter 3

Unraveling the Role of Graphene Nanosheets in Electric Discharge Machining

Kamlesh Paswan, Somnath Chattopadhyaya and Anil Dube

Abstract

The role of graphene nanosheets shows a promising advantage over conventional machining processes. The preparation and characteristics of graphene nanosheets mixed dielectric are investigated with the help of specialized equipment. Also, the machining performance is compared at different concentrations of graphene nanosheets in the dielectric medium with conventional electrical discharge machining. Cetyl trimethyl ammonium bromide (CTAB) and sodium dodecyl sulfate (SDS) surfactants are used for suspension and machining efficiency characterization. The addition of graphene nanosheets into deionized water improves the machining performance. It gives better results at 0.2 g per 800 ml of deionized water. Material removal rate (MRR) improves by 21.27% at lower discharge energy and 114.28% at higher discharge energy. Surface roughness improves by 18% at lower discharge energy and higher spark gap. Using surfactants in the dielectric medium improves the suspension capability of graphene nanosheets in the dielectric medium. However, it reduced the machining efficiency. A slight variation or negligible variation is observed in the density and viscosity of the dielectric medium after adding graphene nanosheets into it.

Keywords: nanosheets, NSMEDM, EDM, graphene, machining

1. Introduction

The dielectric medium influences the electrical discharge machining (EDM) process to a great extent. It acts as a medium for spark generation, cooling the machined surface, and washing debris from the machined surfaces. Several researchers have investigated the role of various dielectric media (deionized water, distilled water, kerosene, air, and oxygen) in EDM [1, 2].

Several investigations reported that impurities in the dielectric significantly affect EDM performance. Adding solid additives in powder into the dielectric medium created a new research area for the graphene Nanosheets Mixed Electrical Discharge Machining Process (NSMEDM) [3, 4]. Further, Kung et al. [5] moved one step ahead and used conductive powder-mixed dielectric medium to machine cobalt-bonded tungsten carbide material through EDM. The powder particles in the dielectric

SI. No.	Ref. no.	Workpiece	Tool	Dielectric medium	Powder (size)
1.	[10]	SKH-54	Cu	EDM oil	Gr (38 ± 3 μm), Si (45 ± 3 μm), Al (45 ± 3 μm), SiC (2.36 ± 0.08 mm), Crushed, glass (2 ± 0.07 mm), MoS_2 (1–3 μm)
Findings	Si and C powders are better for obtaining a fine finish. Gr and Si powders distributed the discharge in the spark gap.				
2.	[11]	EN-31	Cu	Kerosene	Si (20–30 μm)
Findings	Silicon powder enhances the MRR and surface roughness.				
3.	[12]	AISI 1049	Cu	EDM oil	Ti (<38 μm)
Findings	Small discharge energy and power density lead to TiC layer deposition over the machined surface. Hardness increased by threefold.				
4.	[13]	AISI H13	Cu	Kerosene	MWCNT
Findings	Average MRR improved by 26.87%. Average RCL reduced by 30.89%.				
5.	[14]	Ti-6Al-4 V	Cu	Kerosene	Gr (37 μm)
Findings	Improved MRR, roughness, and RLT. Stable machining.				
6.	[15]	Inconel 625	Cu	Kerosene	Gr (15 μm)
Findings	Surface roughness improved—larger crater size. Surface crack density reduced. Reduction in RLT. Carbone formation due to increased pyrolysis.				
7.	[16]	Ti-6Al-4 V	Cu	EDM oil	SiC
Findings	RLT increases with an increase in peak current. Hardness increases by twofold.				
8.	[17]	Ti-6Al-4 V	Cu	Oil Flux ELF2	MWCNT
Findings	Better machinability is achieved at long-pulse on time and low-energy pulse. MRR is higher at long-pulse, which is associated with low current. Reduces the size of micro-surface cracks.				
9.	[18]	Ti-6Al-4 V	Cu	EDM oil + Span 20	Gr (30 nm)
Findings	Surfactant reduces the agglomeration effect. Surfactants also reduce RLT and microcracks.				
10.	[19]	Inconel 706	Cu	EDM oil	Al (1 μm)
Findings	109% MRR increases by changing pulse on-time from 1 μs to 6 μs and 95% when peak current changes from 3 A to 70 A, 25.9% surface roughness, and 13.7% TWR increases when peak current changes from 3A to 70A.				
11.	[20]	NAK80 steel	Cu	kerosene	Si, Al, Gr, CNT
Findings	Roughness improves by 70%, and machining efficiency improves by 66% with CNT powder.				

Table 1.
Powder used in EDM and findings.

medium dispersed the discharge energy and enhanced the machining efficiency to a great extent. The powder concentration in the dielectric medium has a significant effect on MRR. The machining performance increases to a specific limit with the powder concentration and starts decreasing due to arcing. The grit size of powders is also one of the significant factors in powder-mixed EDM. So, Syed et al. [6] tested the electrical discharge machining (EDM) of W200 Die steel. They used both pure distilled water and a dielectric medium consisting of distilled water with suspended powder. The findings indicate that the material removal rate (MRR) increases as the grit size rises up to 27 μm but then decreases [7, 8]. Kurafuji et al. [9] were the first to employ a mixed powder dielectric in the EDM process. Numerous studies have since explored the addition of various powders to the dielectric medium, and the advantages are summarized in **Table 1**.

The graphene nanosheets' selection depends on the dielectric type and workpiece material. There are two types of graphene nanosheets used in the EDM process:

Conductive: The addition of the conductive graphene nanosheets reduces the insulating strength of the dielectric medium. Therefore, the interelectrode gap/spark gap increases, giving away the easy removal of debris from the machined zone. When applying a potential difference, ample positive and negative charges form on the graphene nanosheets. Electric field intensity was higher at the nearest point between the tool and the workpiece. When the dielectric breakdown occurs at the closest point, the discharge channel is formed between the graphene nanosheets through the tool and the workpiece. The graphene nanosheets distributed the discharge over the workpiece, which widened the discharge gap. It also increases electric field intensity three times higher than conventional EDM [21]. Researchers have used many conductive graphene nanosheets in EDM, such as Al [22, 23], Gr [24, 25], Si [8, 26, 27], Cu [28], SiC [29, 30], Cr [31], B_4C [32], and W [33].

Nonconductive: Adding nonconductive nanosheets into the dielectric medium increases the electric field intensity by 1.5 times higher than conventional EDM. Therefore, the electrode gap increases by one and a half times more than in traditional EDM. Response parameters such as higher MRR and lower surface roughness were achieved compared to the conventional EDM. The EDM process has used the nonconductive powders Al_2O_3 and SiO [34, 35].

Graphene nanosheets have distinct properties from other nanosheets. It has 1500–2500 W/mK thermal conductivity at room temperature, twice that of a diamond. Electrical conductivity is 10^4–10^5 S/m, among the highest of any known material at room temperature, which is 13 time that of copper. The thickness of the single-layer graphene could be equal to only one atom thick, about 0.335 nm. Therefore, it is also called 2D.

2. Experimental section

The graphene nanosheet dimension emphasized an essential parameter in NSMEDM to obtain the desired results. The size of graphene nanosheets affects spark gap, surface roughness, MRR, and tool wear rate (TWR). An increase in nanosheet size increases the spark gap, whereas maximum MRR is achieved in smaller graphene nanosheets. However, TWR followed the reverse trend than MRR [34, 35]. Additive size also affects machined surface quality. The smallest nanosheet size gives a better surface finish. An Amine functionalized graphene nanosheet, called graphene nanosheets, is used in this experiment. Specification of the Amine functionalized graphene nanosheets is given in **Table 2**. The composition and size of graphene nanosheets are determined using a scanning electron microscope (SEM) and energy dispersive X-ray spectroscopy (EDX) [36] equipment, shown in **Figure 1**.

The graphene nanofluid is prepared by mixing graphene nanosheets into deionized water. First, a conical flask consisting of 800 ml of deionized water and a magnetic stirrer machine (Tarson Digital SPINNOT) stir the fluid [37]. After that, a

Thickness (Z)	5–10 nm
Dimension (X&Y)	5–10 μm
NH_2 Ratio	2–5%
Purity	99%
Bulk Density	0.45 g/cm^3
Number of layers	5–10

Table 2.
Technical specification of amine functionalized graphene nanosheets.

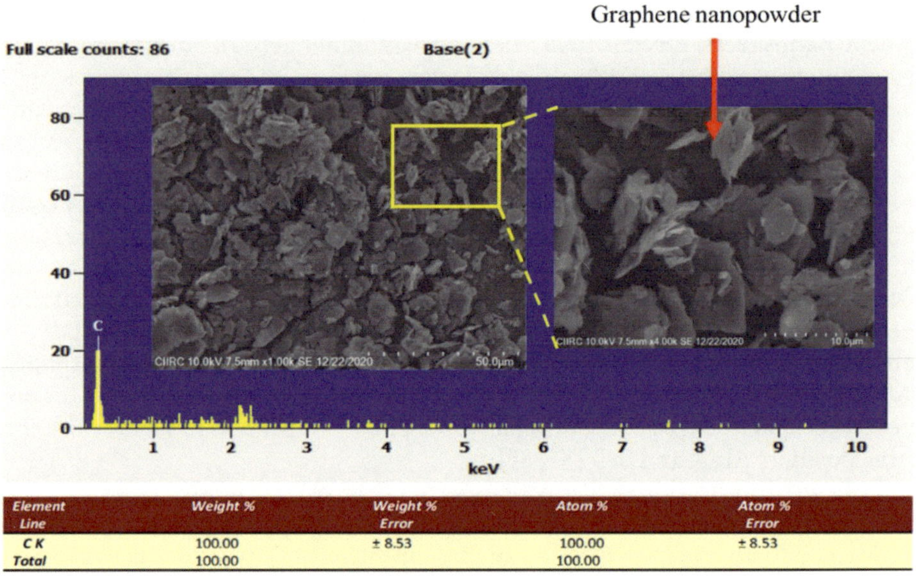

Figure 1.
SEM and EDX result of amine functionalized graphene nanosheets.

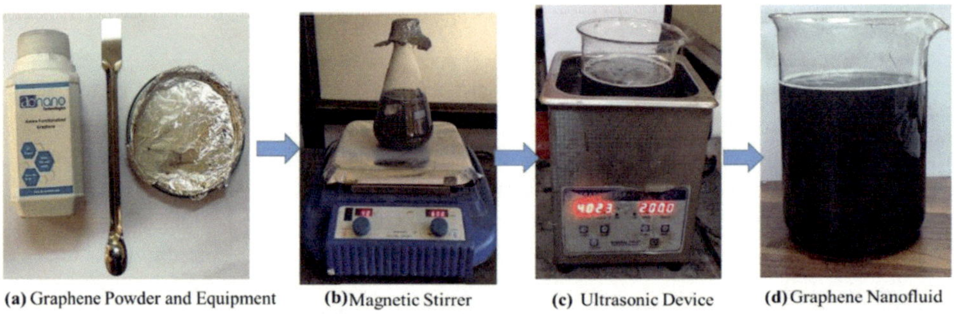

Figure 2.
Graphene nanosheets mixed dielectric preparation.

small amount of graphene nanosheets were poured into the conical flask with an average interval of 1–2 min. After 5 hours of stirring, the nanofluid is ultrasonicated in an ultrasonic device for an hour. **Figure 2** shows the process of nanofluid preparation.

3. Results and discussion

The concentration of graphene nanosheets in NSMEDM plays a vital role. Abnormal discharge occurs mainly in conventional EDM with a slight chance of short-circuiting. With the increase in nanosheet concentration in the dielectric medium, abnormal discharge reduces, and at a specific concentration, abnormal discharge is eliminated [38]. So, the dielectric gives maximum machining efficiency at a particular concentration range. On further addition of graphene nanosheets into the dielectric, the chances of short-circuiting increase by which reduction in the machine efficiency started. The excess amount of graphene nanosheets bridges the tool and the workpiece, leading to short-circuiting and abnormal discharge [39]. **Figure 3** shows the effect of different graphene concentrations [no graphene (0), 0.1, 0.2, 0.4, and 0.6] in 800 ml of deionized water. It shows that at 0.2 g/800 ml, nanofluid gives better MRR. It improves results from 21.27 to 114.28% at the 0.2 g/800 ml concentration. The machined surface Ra, Rq, and Rz also improved by 18, 21, and 28%. The plot of surface roughness at different concentrations is shown in **Figure 4**.

One of the significant issues in NSMEDM is the suspension of graphene nanosheets into the dielectric medium. It was found that after a few days of suspension, graphene nanosheets settled down and were agglomerated at the bottom because of the surface tension between the graphene nanosheets and dielectric fluid. Surfactant is more likely detergents that reduce the surface tension between the solid and liquid. This allows graphene nanosheets to float for a longer time, and even after the settlement at the bottom, graphene nanosheets are kept evenly distributed at the bottom. The preliminary investigation shows that without surfactant, the suspension

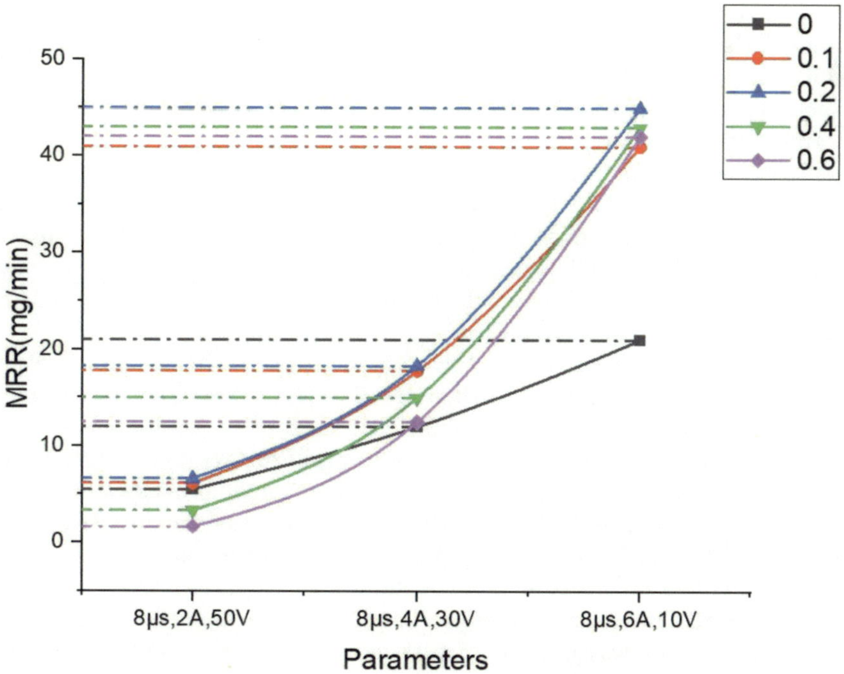

Figure 3.
Effect of concentration on MRR at different parameters and concentration of graphene nanosheets.

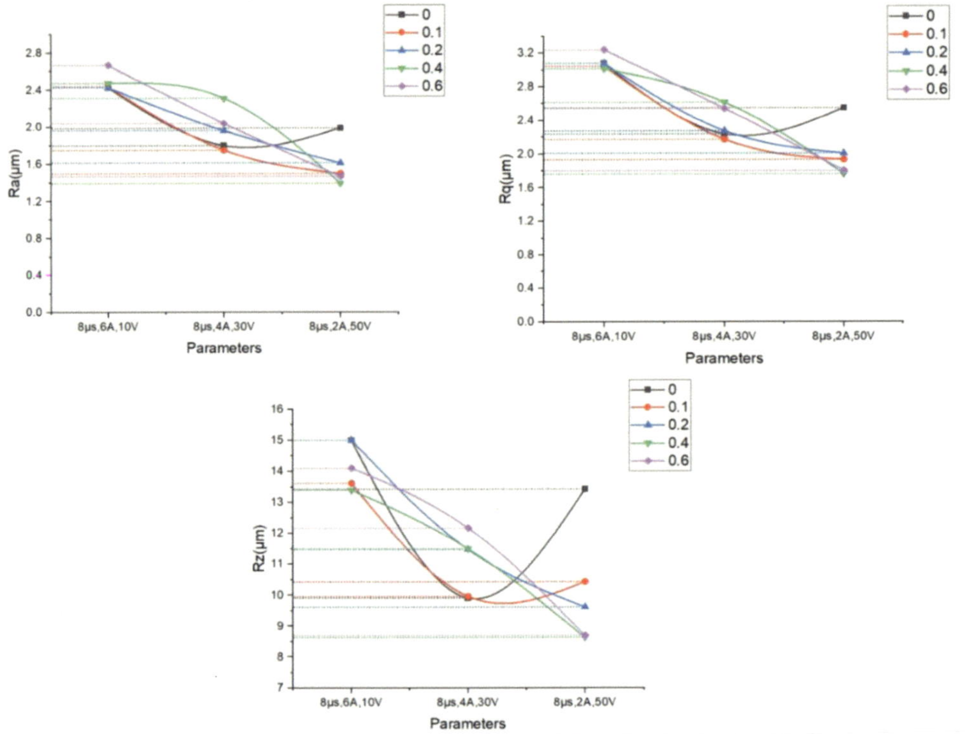

Figure 4.
Effect of concentration on surface roughness at different parameters and concentration of graphene nanosheets.

of graphene lasts for 2–3 hours, and the addition of surfactant increases the suspension, which can last more than a month [40].

Surfactants are widely used during the preparation of nanofluids. It increases the suspension period of the graphene nanosheets in the medium. This investigation uses SDS and CTAB with 0.2 g graphene in 800 ml deionized water. Surfactants are soluble and mix entirely into the dielectric medium, increasing the electrical conductivity. **Figure 5** shows the suspension time for pure graphene and graphene with surfactant in deionized water. The electrical conductivity of the nanosheet concentration was measured, and deionized has a conductivity of 6 μs/cm, increasing up to 24 μs/cm after adding 0.2 g graphene nanosheets.

Further, the addition of surfactant also increases the electrical conductivity by adding 0.1 g SDS, 0.2 g SDS, 0.1 g CTAB, and 0.2 g CTAB separately, increasing electrical conductivity from 24 to 76 μs/cm, 125, 116, and 165 μs/cm, respectively. **Figure 6** shows the effect of graphene nano power and surfactant on MRR. The graph shows that the highest MRR obtained at 0.2 g graphene in deionized water without any surfactant is 45 mg/min. Whereas MRR obtained with surfactants SDS at 0.1 and 0.2 g and CTAB at 0.1 and 0.2 g are 31.81 and 43 mg/min, and 28.33 and 30 mg/min, respectively. The surface roughness of Ra, Rq, and Rz is improved by adding surfactants. **Table 3** shows the response values of surface roughness with varying surfactants. **Figure 7** shows the effect of surfactant at different conditions on surface roughness. The dielectric medium with graphene nanosheets without surfactant gives

Figure 5.
Suspension of graphene nanofluid.

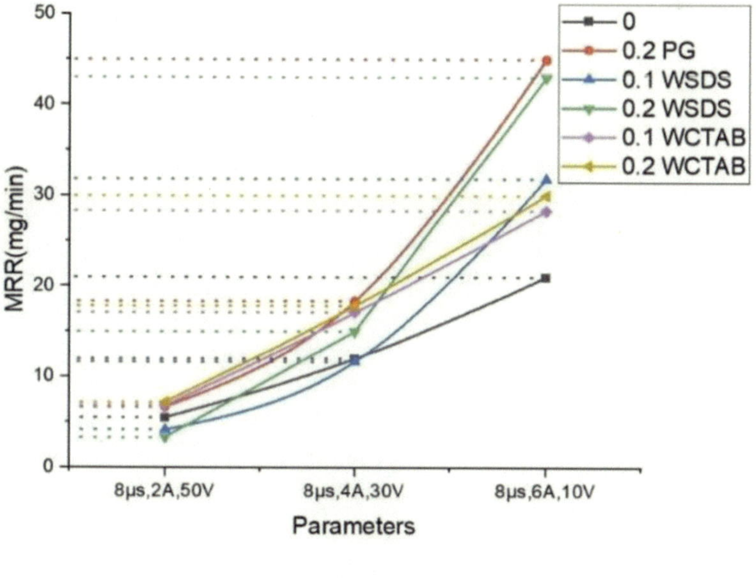

Figure 6.
MRR response at (0) no graphene, (0.2PG) 0.2 g nanographene, 0.2 g graphene with 0.1 and 0.2 g SDS and CTAB in a dielectric, respectively.

	Input parameters	0 No graphene	0.2 g graphene	0.1 g SDS	0.2 g SDS	0.1 g CTAB	0.2 g CTAB
Ra	8 μs, 2 A, 50 V	1.99	1.61	2.41	1.92	1.69	1.94
	8 μs, 4 A, 30 V	1.79	1.96	2.11	2.02	1.79	2.08
	8 μs, 6 A, 10 V	2.43	2.43	2.3	2.11	2.41	2.44
Rq	8 μs, 2 A, 50 V	2.55	2.01	2.99	2.45	2.12	2.38
	8 μs, 4 A, 30 V	2.24	2.27	2.58	2.58	2.21	2.62
	8 μs, 6 A, 10 V	3.08	3.08	2.87	2.67	3.04	3.21
Rz	8 μs, 2 A, 50 V	13.42	9.61	13.4	11.53	9.93	12.00
	8 μs, 4 A, 30 V	9.9	11.47	11.84	12.02	10.60	12.94
	8 μs, 6 A, 10 V	15	15	13.82	12.52	15.06	14.11

Table 3.
Surface roughness response values for SDS and CTAB surfactant.

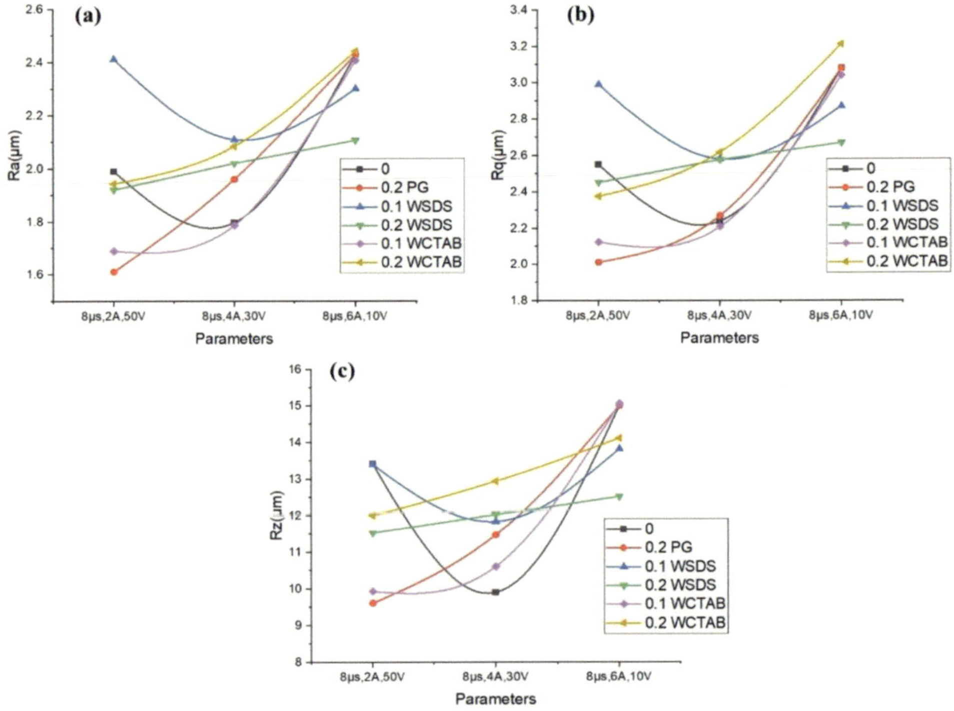

Figure 7.
Roughness response values at no graphene, 0.1 and 0.2 g SDS and CTAB in graphene nanosheets mixed dielectric.

better results than those without nanosheets and dielectric with graphene nanosheets and surfactant.

The characterization is based on the physical properties of the graphene nanosheets mixed with dielectric fluid, such as density, conductivity, and viscosity.

Density: A hydrometer (500 ml) with a measuring cylinder to measure the dielectric medium's density with and without nanosheets. There is no difference in the

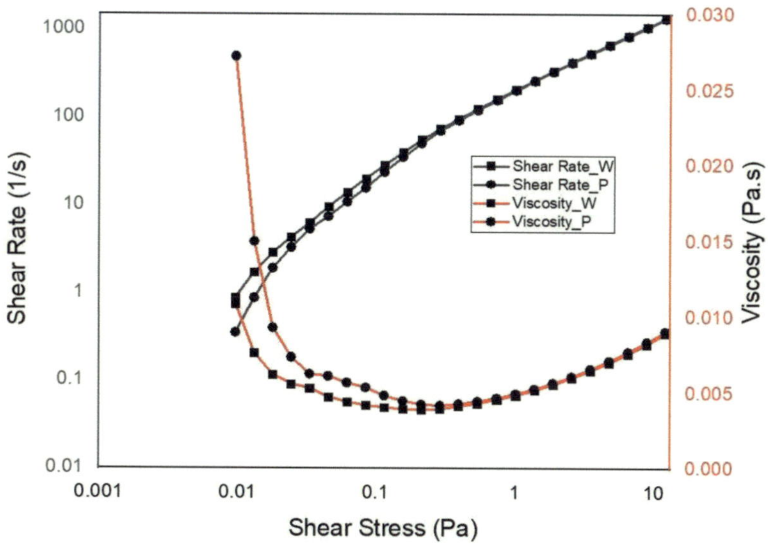

Figure 8.
Viscosity, shear rate, and shear stress plot for the dielectric medium with and without nanosheets.

density as a result obtained is 1000 kg/m^3,which is identical in both the cases with and without nanosheets.

Electrical conductivity: The electrical conductivity of the dielectric medium with and without nanosheets is measured using a Hanna conductivity meter (Model HI 991300) and found that deionized without nanosheets has conductivity 6 μs/cm, increasing up to 24 μs/cm after adding 0.2 g graphene nanosheets into the deionized.

Viscosity test: Various ways to measure the viscosity of a fluid. Malvern Instruments made a viscometry (Bohlin Gemini 2) in the UK to measure the dielectric medium's viscosity. **Figure 8** shows the dielectric medium's viscosity without nanosheets at room temperature (25°C). It is noticed that the viscosity of the dielectric medium increases with the addition of graphene nanosheets. The viscosity difference is high at lower shear stress and lowers at higher shear stress. The dielectric medium's viscosity without nanosheets and graphene nanosheets at shear stress 0.007 Pa are 0.01086 and 0.02721 Pa.s, respectively.

4. Conclusion

This experiment discusses the preparation and characteristics of graphene nanosheets mixed dielectric. Also, the machining performance is compared at different concentrations of graphene nanosheets in the dielectric medium with conventional EDM. CTAB and SDS surfactants are used for suspension and machining efficiency characterization.

- The addition of graphene nanosheets into deionized water improves the machining performance. It gives better results at 0.2 g per 800 ml of deionized water.

- MRR improves by 21.27% at lower discharge energy and 114.28% at higher discharge energy.

- Surface roughness improves by 18% at lower discharge energy and higher spark gap.
- Using surfactants in the dielectric medium improves the suspension capability of graphene nanosheets in the dielectric medium. However, it reduced the machining efficiency.
- A slight variation or negligible variation is observed in the density and viscosity of the dielectric medium after adding graphene nanosheets into it.

Author details

Kamlesh Paswan[1*], Somnath Chattopadhyaya[2] and Anil Dube[1]

1 Sandip Institute of Engineering and Management, Nashik, India

2 Indian Institute of Technology (Indian School of Mines), Dhanbad, India

*Address all correspondence to: kamleshpgcr@gmail.com

References

[1] Paswan K, Chattopadhyaya S, Pramanik A. A review paper on machining of metal matrix composite and optimizing methods used in electrical discharge machining. Materials Today: Proceedings. 2018;**5**:24428-24438. DOI: 10.1016/j.matpr.2018.10.239

[2] Paswan K, Ranjan R, Hembram S, Bajaj R, Rai Dixit A, Mandol A, et al. Experimental investigation on machining performance using deionized water as dielectric in die-sinking EDM. Indian Journal of Science and Technology. 2016;**9**:2-6. DOI: 10.17485/ijst/2016/v9i34/100912

[3] Erden A, Bilgin S. Role of impurities in electric discharge machining. In: Proc Twenty-Fourth Int Mach Tool des Res Conf. Vol. 21. 1980. pp. 345-350. DOI: 10.1007/978-1-349-05861-7_45

[4] Jeswani ML. Effect of the addition of graphite powder to kerosene used as the dielectric fluid in electrical discharge machining. Wear. 1981;**70**:133-139. DOI: 10.1016/0043-1648(81)90148-4

[5] Kung KY, Horng JT, Chiang KT. Material removal rate and electrode wear ratio study on the powder mixed electrical discharge machining of cobalt-bonded tungsten carbide. International Journal of Advanced Manufacturing Technology. 2009;**40**:95-104. DOI: 10.1007/s00170-007-1307-2

[6] Syed KH, Kuppan P. Studies on recast-layer in EDM using aluminium powder mixed distilled water dielectric fluid. International Journal of Engineering and Technology. 2013;**5**:1775-1780

[7] Laxmi D, Paswan K, Chattopadhyaya S, Pramanik A. Influence of low-frequency vibration in die sinking EDM: A review influence of low-frequency vibration in die sinking EDM: A review. IOP Conference Series: Materials Science and Engineering. 2021;**1104**:012010. DOI: 10.1088/1757-899X/1104/1/012010

[8] Paswan K, Pratap Singh S, Chattopadhyaya S, Pramanik A. Experimental investigation on the effects of aqueous solution in electric discharge machining. Materials Today: Proceedings. 2020;**27**:2975-2980. DOI: 10.1016/j.matpr.2020.04.904

[9] Kurafuji H, Suda K. Study on electrical discharge machining I. Journal of the Faculty of Engineering, University of Tokyo. Series B. 1965;**XXVIII**:1-18

[10] Wong Y, Lim L, Rahuman I, Tee W. Near-mirror-finish phenomenon in EDM using powder-mixed dielectric. Journal of Materials Processing Technology. 1998;**79**:30-40. DOI: 10.1016/S0924-0136(97)00450-0

[11] Kansal HK, Singh S, Kumar P. Parametric optimization of powder mixed electrical discharge machining by response surface methodology. Journal of Materials Processing Technology. 2005;**169**:427-436. DOI: 10.1016/j.jmatprotec.2005.03.028

[12] Furutani K, Sato H, Suzuki M. Influence of electrical conditions on performance of electrical discharge machining with powder suspended in working oil for titanium carbide deposition process. International Journal of Advanced Manufacturing Technology. 2009;**40**:1093-1101. DOI: 10.1007/s00170-008-1420-x

[13] Mohammadzadeh Sari M, Noordin MY, Brusa E. Role of multi-wall carbon nanotubes on the main

parameters of the electrical discharge machining (EDM) process. International Journal of Advanced Manufacturing Technology. 2013;**68**:1095-1102. DOI: 10.1007/s00170-013-4901-5

[14] Unses E, Cogun C. Improvement of electric discharge machining (EDM) performance of Ti-6Al-4V alloy with added graphite powder to dielectric. Journal of Mechanical Engineering. 2015;**61**:409-418. DOI: 10.5545/sv-jme.2015.2460

[15] Talla G, Gangopadhyay S, Biswas CK. Influence of graphite powder mixed EDM on the surface integrity characteristics of inconel 625. Particulate Science and Technology. 2017;**35**:219-226. DOI: 10.1080/02726351.2016.1150371

[16] Li L, Zhao L, Li ZY, Feng L, Bai X. Surface characteristics of Ti-6Al-4V by SiC abrasive-mixed EDM with magnetic stirring. Materials and Manufacturing Processes. 2017;**32**:83-86. DOI: 10.1080/10426914.2016.1151043

[17] Shabgard M, Khosrozadeh B. Investigation of carbon nanotube added dielectric on the surface characteristics and machining performance of Ti–6Al–4V alloy in EDM process. Journal of Manufacturing Processes. 2017;**25**:212-219. DOI: 10.1016/j.jmapro.2016.11.016

[18] Kolli M, Kumar A. Surfactant and graphite powder-assisted electrical discharge machining of titanium alloy. Proceedings of the Institution of Mechanical Engineers Part B Journal of Engineering Manufacture. 2017;**231**:641-657. DOI: 10.1177/0954405415579019

[19] Wang C, Qiang Z. Comparison of micro-EDM characteristics of Inconel 706 between EDM oil and an Al powder-mixed dielectric. Advances in Materials Science and Engineering. 2019;**2019**:1-11. DOI: 10.1155/2019/5625360

[20] Mai C, Hocheng H, Huang S. Advantages of carbon nanotubes in electrical discharge machining. International Journal of Advanced Manufacturing Technology. 2012;**59**:111-117. DOI: 10.1007/s00170-011-3476-2

[21] Kozak J, Rozenek M, Dabrowski L. Study of electrical discharge machining using powder-suspended working media. Proceedings of the Institution of Mechanical Engineers Part B Journal of Engineering Manufacture. 2003;**217**:1597-1602. DOI: 10.1243/095440503771909971

[22] Chow HM, Yan BH, Huang FY, Hung JC. Study of added powder in kerosene for the micro-slit machining of titanium alloy using electro-discharge machining. Journal of Materials Processing Technology. 2000;**101**:95-103. DOI: 10.1016/S0924-0136(99)00458-6

[23] Yih-Fong T, Fu-Chen C. Investigation into some surface characteristics of electrical discharge machined SKD-11 using powder-suspension dielectric oil. Journal of Materials Processing Technology. 2005;**170**:385-391. DOI: 10.1016/j.jmatprotec.2005.06.006

[24] Kumar A, Maheshwari S, Sharma C, Beri N. Machining efficiency evaluation of cryogenically treated copper electrode in additive mixed EDM. Materials and Manufacturing Processes. 2012;**27**:1051-1058. DOI: 10.1080/10426914.2011.654151

[25] Liew PJ, Yan J, Kuriyagawa T. Carbon nanofiber assisted micro electro discharge machining of reaction-bonded silicon carbide. Journal of Materials Processing Technology. 2013;**213**:1076-1087. DOI: 10.1016/j.jmatprotec.2013.02.004

[26] Molinetti A, Amorim FL, Soares PC, Czelusniak T. Surface modification of AISI H13 tool steel with silicon

or manganese powders mixed to the dielectric in electrical discharge machining process. International Journal of Advanced Manufacturing Technology. 2016;**83**:1057-1068. DOI: 10.1007/s00170-015-7613-1

[27] Henriques E, Pecas P. Influence of silicon powder-mixed dielectric on conventional electrical discharge machining. International Journal of Machine Tools and Manufacture. 2003;**43**:1465-1471. DOI: 10.1016/S0890-6955(03)00169-X

[28] Kansal HK, Singh S, Kumar P, Engineering M. Effect of silicon powder mixed EDM on machining rate of AISI D2 die steel. Journal of Manufacturing Processes. 2007;**9**:13-22. DOI: 10.1016/S1526-6125(07)70104-4

[29] Chow HM, Yang LD, Lin CT, Chen YF. The use of SiC powder in water as dielectric for micro-slit EDM machining. Journal of Materials Processing Technology. 2008;**195**:160-170. DOI: 10.1016/j.jmatprotec.2007.04.130

[30] Ekmekci B, Ersöz Y. How suspended particles affect surface morphology in powder mixed electrical discharge machining (PMEDM). Metallurgical and Materials Transactions B: Process Metallurgy and Materials Processing Science. 2012;**43**:1138-1148. DOI: 10.1007/s11663-012-9700-0

[31] Kumar S, Dhingra AK, Kumar S. Parametric optimization of powder mixed electrical discharge machining for nickel-based superalloy inconel-800 using response surface methodology. Mechanics of Advanced Materials and Modern Processes. 2017;**3**:1-17. DOI: 10.1186/s40759-017-0022-4

[32] Bhattacharya A, Batish A, Kumar N. Surface characterization and material migration during surface modification of die steels with silicon, graphite and tungsten powder in EDM process. Journal of Mechanical Science and Technology. 2013;**27**:133-140. DOI: 10.1007/s12206-012-0883-8

[33] Furutania K, Saneto A, Takezawa H, Mohri N, Miyake H. Accretion of titanium carbide by electrical discharge machining with powder suspended in working fluid. Precision Engineering. 2001;**25**:138-144. DOI: 10.1016/S0141-6359(00)00068-4

[34] Tzeng YF, Lee CY. Effects of powder characteristics on electrodischarge machining efficiency. International Journal of Advanced Manufacturing Technology. 2001;**17**:586-592. DOI: 10.1007/s001700170142

[35] Jabbaripour B, Sadeghi MH, Shabgard MR, Faraji H. Investigating surface roughness, material removal rate and corrosion resistance in PMEDM of γ-TiAl intermetallic. Journal of Manufacturing Processes. 2013;**15**:158-166. DOI: 10.1016/j.jmapro.2012.09.016

[36] Krishnamoorthy K, Veerapandian M, Zhang L-H, Yun K, Kim SJ. Antibacterial efficiency of graphene nanosheets against pathogenic bacteria via lipid peroxidation. Journal of Physical Chemistry C. 2012;**116**:17280-17287. DOI: 10.1021/jp3047054

[37] Paswan K, Chattopadhyaya S, Pramanik A. Performance of graphene nanopowder with deionised water in EDM process. Materials Science Forum. 2021;**1026**:147-154. DOI: 10.4028/www.scientific.net/MSF.1026.147

[38] Peças P, Henriques E. Effect of the powder concentration and dielectric flow in the surface morphology in electrical discharge machining with powder-mixed dielectric

(PMD-EDM). International Journal of Advanced Manufacturing Technology. 2008;**37**:1120-1132. DOI: 10.1007/s00170-007-1061-5

[39] Reddy VV, Kumar A, Valli PM, Reddy CS. Influence of surfactant and graphite powder concentration on electrical discharge machining of PH17-4 stainless steel. Journal of the Brazilian Society of Mechanical Sciences and Engineering. 2015;**37**:641-655. DOI: 10.1007/s40430-014-0193-4

[40] Paswan K, Pramanik A, Chattopadhyaya S. Machining performance of Inconel 718 using graphene nanofluid in EDM. Materials and Manufacturing Processes. 2020;**35**:33-42. DOI: 10.1080/10426914.2020.1711924

Chapter 4

Heavy Metals Adsorption by Nanosheet: Mechanism and Effective Parameters

Mostafa Khosroupour Arabi and Morteza Ghorbanzadeh Ahangari

Abstract

Among the methods for heavy metals removal such as precipitation, evaporation, electroplating and ion exchange, which have many disadvantages, adsorption is the cost effective and environmental friendly technique. Using nanosheets as the base materials for the adsorption because of their large surface area and high adsorption capacity is broadened. Carbon products (Graphene), boron nitride materials (BNM), transition metal dichalcogenides (TMDs), layered double hydroxiades (LDHs) and MXene are most well-known nanosheets, which have been used for heavy metal ions removal from aqueous solutions. In this review, experimental and simulation studies on nanosheet adsorbents are presented to pinpoint the importance of this group of nano-materials on water/wastewater treatment technology. Molecular dynamics (MD) and density functional theory (DFT) are the most common simulation methods for demonstration of adsorption mechanism of nanosheets. In addition, synthesis methods, adsorption mechanism, adsorption performance, and effective parameters of nanosheets and novel techniques to improve the adsorption capability and regeneration of adsorbents are introduced. The adsorption of heavy metals to nanosheet mechanisms is also introduced and isotherm/ kinetics models for each nanosheet are evaluated. With all the advantages of nanosheets, it should be noted that their usage in larger industrial scales should be further investigated.

Keywords: adsorption, heavy metals removal, nanosheets, effective parameters, experimental and simulation analysis

1. Introduction

Replace the entirety of this text with the introduction to your chapter. The introduction section should provide a context for your manuscript and should be numbered as first heading. When preparing the introduction, please bear in mind that some readers will not be experts in your field of research.

From the time when nanosheets were introduced to scientific society, scientists have been achieving more properties and applications of these nanoparticles in every aspect of science such as engineering, environment, chemistry, physics, and so on [1–4]. Strong adsorption capability of nanosheets are due to their high surface area, porosity, specific surface charge, surface functionality, and ions adsorption characteristics [5–8] originating from their small size structures have drawn the attention for energy and water and wastewater treatment applications [9–12].

Today, expansion of various industries including mining industry, sewage irrigation, metal plating, and electronic industry has contaminated the water and wastewater with heavy metals such as cadmium (Cd), cobalt (Co), chromium (Cr), copper (Cu), and mercury (Hg) [13]. Human health and the ecosystem are in danger as a result of the heavy metal entering the environment. The presence of heavy metal in water in excess could cause serious health problems for living thing because they are non-biodegradable and may be carcinogenic [14]. As a result, various methods for removing heavy metal ions have been developed, including coagulation and flocculation, precipitation, ion exchange, membrane technologies, electrochemical technologies, and adsorption [15, 16]. There are only a few treatment methods that are used due to various economic and technological factors. These processes have some drawbacks (high cost, process complexity, low efficiency, etc.), despite the fact that they can remove various pollutants from water and wastewaters. It is interesting to note that the adsorption method, which is regarded as the safest, cleanest, and most practical and technically feasible process, was deemed to be the most effective for removing heavy metals [17]. Various adsorbents, such as fly ash [18, 19], activated carbon [20, 21], zeolite [22–24], montmorillonite [25–27], and chitosan [28–30], have been reported for heavy metal removal from aqueous solutions. Nanomaterials frequently have unique characteristics under the nanoscale, for instance a surface effect, small-scale impact, quantum effect, and macro-quantum tunnel effect in contribution with their extraordinary adsorption capacity and reactivity cause them to be the proper materials for removal of heavy ions [31, 32]. A new class of nanomaterials with sheet-like morphologies called 2D nanosheets is made up of layer structures with lateral sizes greater than 100 nm and thickness smaller than 10 nanometers. Physical and chemical characteristics of nanosheets are dependent on formation of their structures from inorganic compounds. Due to their mechanical strength, flexible structure, chemical inertness, and separation performance, nanosheets have shown promise as building blocks or fillers for membranes [33, 34].

Carbon products (graphene), boron nitride materials (BNM), transition metal dichalcogenides (TMDs), layered double hydroxides (LDHs), and MXene are most common 2D nanosheets [35] and some novel nanosheets [36–45] implemented in water and wastewater treatment. The novel carbon nanomaterial adsorbents include carbon nanotubes, graphene, and other nano-adsorbents with carbon element and sp2 as the main hybrid form. Graphene has a two-dimensional structure, which is the foundation for creating graphitic materials in all dimensions. For instance, it can be folded into on-dimensional carbon nanotubes, enveloped in zero-dimensional fullerene, or layered into three-dimensional graphite. Its crystal lattice is composed of a single layer crystalline carbon atoms and has a hexagonal lattice shape. Strong adsorption capacity, high removal efficiency, fast equilibrium speed, low cost, no secondary pollution, and good recyclability are only a few of the notable benefits of using graphene as an adsorbent to remove heavy metals [46]. There are typically four main forms of crystalline BNMs, including hexagonal/rhombohedral BNMs that resemble graphite and have dense phases with sp2 hybridized B-N bonds and cubic BNMs that resemble wurtzite and diamond and have low density phases with sp3 hybridized links. Its several physical forms of BN nanomaterials are 0D, 1D, 2D, and 3D shapes like nanoparticles/fullerenes, nanotubes/nanofibers/nanoribbons, nanosheets/nanomeshes, and abd nanoflowers/hollow spheres, respectively. Particularly, BNMs and BN-based nanomaterials have been thoroughly studied for possible environmental applications such as adsorption, photocatalytic degradation, and membrane separation to remove pollutants [47]. TMDs, including MoS_2, WS_2, $MoSe_2$, and WSe_2, have been known as semiconducting 2D layered materials. The elements are described as MX_2, where M is a transition metal form (Group 4 to 10), and X stands for chalcogen elements of S, Se, or Te, which can crystallize in either a non-layer form or a layered structure resembling graphite (mainly the Groups 4–7 TMDs). Numerous effective methods, including heteroatom doping, vacancy engineering, and the addition of a second component into or onto 2D TMD nanosheets, have been proposed to further explore the possibilities of these materials. Due to their extensive nanocapillary channels, TMDs nanosheet-based membranes have demonstrated exceptional performance for nanofiltration and desalination in recent years. TMD membranes, as stated, offer more benefits than other 2D nanosheets, such as high water flux and water stability. Additionally, the laminar TMD membranes in water have free spacing that is about sub -1 nm, which aids in molecular separation [48, 49]. LDHs are a broad category of positively charged metal hydroxide layers that resemble brucite and charge-balancing intercalated anions. LDHs have substantial specific surface areas and strong ion and molecule adsorption capacities, much like other two-dimensional materials. Inorganic metal hydroxide layers' chemical composition, structure, size, shape, and interlayer gallery anions all affect how well LDHs adsorb substances. The chemical identity of the metal ions and the size of the intercalated anions govern the cationic layer thickness, which in turn controls the chemical reactivity of LDHs. The application of original LDHs in sensing and adsorption are constrained by the functional groups and simple structural components, despite the low cost and simplicity of their preparation [50–52]. Due to their exceptional properties, MXenes is one of the newest and most adaptable two-dimensional; nanomaterials have attracted a lot of attention in recent years. One of the four primary structures M_2AX, M_3, M_4, M_5AX_4, where M is an early transition metal group, A is a group A element (e.g., Al, Si, or Ga), and X is carbon and/or nitrogen, produces MXenes, which are conversion metal carbides, nitrides, and carbonicides from densely layered ceramic precursors known as MAX phases. MAX phases themselves have been used in water treatment applications because of their excellent environmental stability, nontoxicity, and simplicity of preparation. However, the selective etching of an element from the MAX phase to produce a colloidal MXene solution with an accordion shape can produce many distinctive surface characteristics, including hydrophilic surfaces, large surface areas, high electronic conductivity, an abundance of surface functional groups, redox properties, and high capacity for absorption. Adsorption of organic and inorganic contaminants, photocatalysis, water treatment membranes, capacitive deionization, and antimicrobial applications are just a few of the many environmental remediation used for MXenes [53].

In present work, a brief review on heavy metal removal using 2D nanosheets is investigated from experimental and simulation aspect of view. Studies are about common and novel nanosheets, which were prepared in laboratories and tested to evaluate and compare the adsorption capability of heavy metal ions from aqueous solutions. In addition, DFT and MD simulations have been done to study the adsorption mechanism of the nanosheets and the effective parameters of process improvement. Finally, a conclusion on technical challenges using nanosheets in current and future is presented.

2. 2D nanosheets used for heavy metal removal

2.1 Carbon products (graphene)

Using a modified Hummers process, a few layered graphene oxide (GO) nanosheets could be made from graphite and utilized as sorbents to remove Cd(II) and Co(II) ions from large volumes of aqueous solutions. The results of tests on effective parameters including pH, ionic strength, and acid on Cd(II) and Co(II) sorption showed that the sorption was significantly influenced by pH and minimally related to ionic strength. Cd(II) and Co(II) sorption were affected by the abundant oxygen-containing functional groups on the surface of graphene oxide nanosheets [54].

The colloidal behavior of GO showed that common cations (Ca^{2+}, Mg^{2+}, Na^{+}, and K^{+}) might destabilize GO solution less aggressively than heavy metal cations (Cr^{3+}, Pb^{2+}, Cu^{2+}, Cd^{2+}, and Ag^{+}) (**Figure 1**). Additionally, heavy metal cations can readily cross the electric double-layer (EDL), attach to GO surface, and subsequently change the surface potential, which is a more effective mechanism for GO aggregation. According to aggregation kinetics, $Cr^{3+} >> Pb^{2+} > Cu^{2+} > Cd^{2+} > Ca^{2+} > Mg^{2+} >> Ag^{+} > K^{+} > Na^{+}$ is the order of destabilizing ability of cations. The destabilizing capacity of metal cations is consistent with their tendency to bind to GO, which is governed by their electronegativity and hydration shell thickness [55]. Sunlight-assisted surface nanostructure tailoring involving controllable morphological and structural alterations of graphene could improve the ion adsorption and filtration. Notably, lighting control could be used to sufficiently adjust the interlayer spacing of single-layer graphene to encourage selective interactions between ions and nanosheets [56].

Due to its hydrophilic nature, GO can be dispersed in aqueous media. As a result, it is challenging to remove it from the solution using conventional separation methods after the adsorption process, which could raise the cost of industrial application and/or result in re-pollution of the treated water. With the use of magnetic technology (**Figure 2**), the issue may be resolved [58]. $MnFe_2O_4$ [57, 59], sulfonated magnetic [60, 61], and Fe_3O_4 [57, 62–66] were added to the graphene oxide adsorbents, making it easier for them to be trapped from the medium with the help of an external magnetic field. To improve effectiveness of heavy metal removals, scientists have added other materials

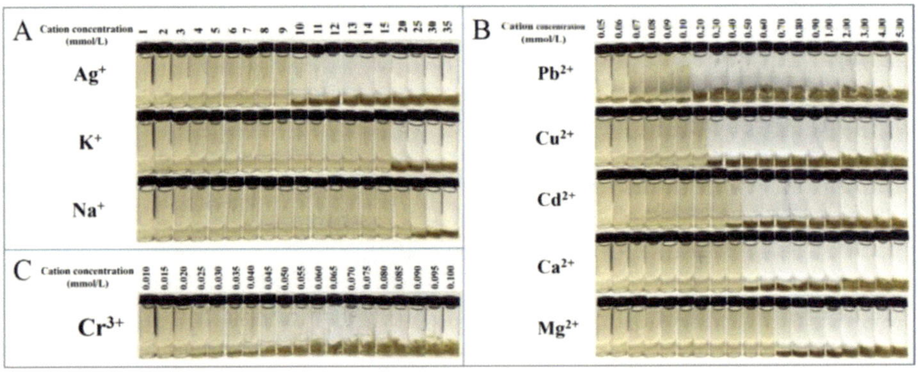

Figure 1.
Bulk flocculation of GO in aqueous solution containing monovalent cations (A), divalent cations (B), and trivalent cations (C) with different concentrations to determine their destabilization capability (https://pubs.acs.org/doi/10.1021/acs.est.6b04235. Note: Further permissions related to the material excerpted should be directed to the ACS) [55].

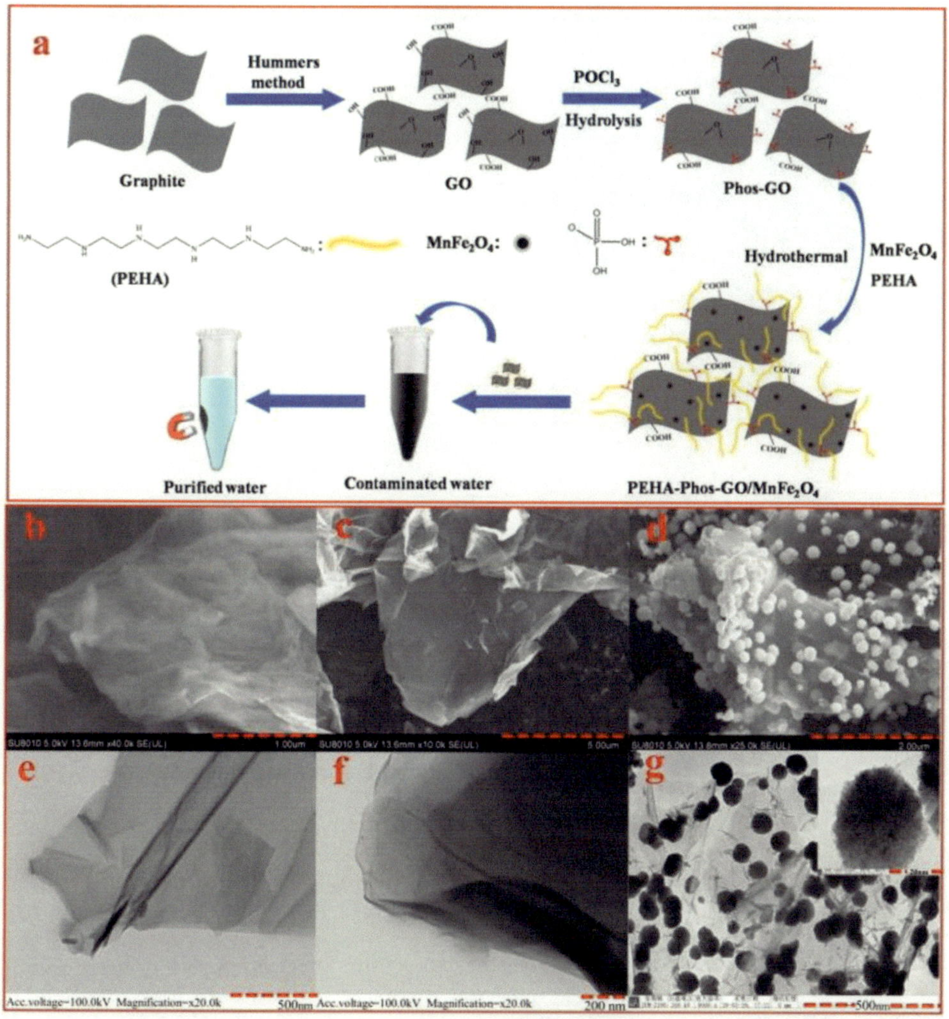

Figure 2.
(a) Schematic illustration of PEHA-Phos-GO/MnFe$_2$O$_4$ nanohybrids preparation and application; SEM images of (b) GO, (c) Phos-GO, and (d) PEHA-Phos-GO/MnFe$_2$O$_4$; TEM images of (e) GO, (f) Phos-GO, and (g) PEHA-Phos-GO/MnFe$_2$O$_4$ [57].

including Chitosan [67–69], Cds [70], silica [71, 72], $Mg(OH)_2$ [73], and CdO [74] to GO nanosheets. Due to its impressive properties, graphydine, a new type of carbon allotrope made up of sp- and sp^2-hybridized carbon atoms, has attracted significant interest. Given that graphydine's acetylenic links interact with metal ions strongly, it is thought to be a promising candidate for the sorption of heavy metals [75–78].

Molecular dynamics (MD) simulations, density functional theory (DFT), and Monte Carlo calculations (MMC) used to investigate the ability of nanoporous graphene (NPG) membranes to desalinate heavy metal ions (**Figure 3**). These studies could give us molecular insight regarding the adsorption sites, interaction type, and adsorption energetics of nanosheets toward heavy metal ions [79]. Nanoporous graphene membrane functionalized with different chemicals (hydroxyl, nitrogen, fluorine, boron, etc.) were studied, and the results showed that NPG functionalized with these groups have high water permeation and ion rejection for different

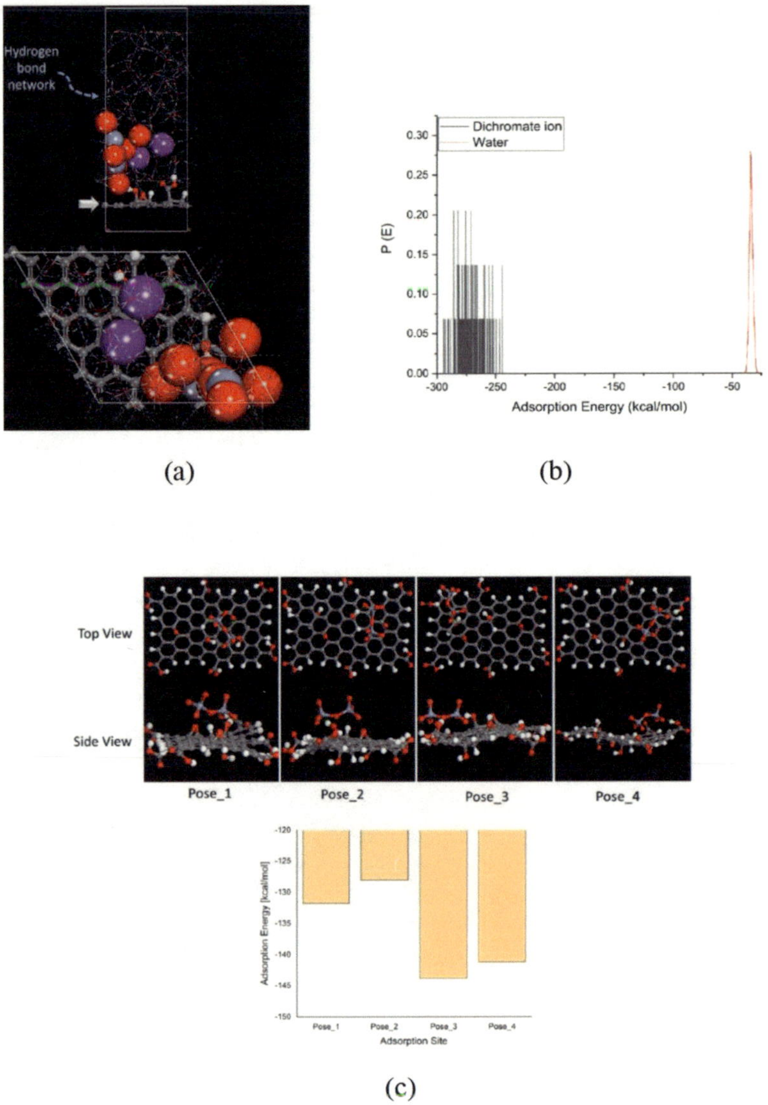

Figure 3.
MD, DFT, and MMC investigation on the ability of nanoporous graphene (NPG) membranes to desalinate heavy metal ions. (a) Side and top views of the equilibrium adsorption configurations of the dichromate ions (one dichromate ion +100 water molecules) onto the GO model (5 × 5 graphene layer containing: An epoxy, carboxy, and hydroxyl functional groups on its surface obtained using MC simulations) (b) adsorption energy distributions for dichromate ions (c) four different explored adsorption sites for the dichromate ion on the GO surface and the corresponding adsorption energy values [79].

conditions [80–83]. MD simulation studied two graphene membranes that are oppositely exposed to two external electric fields used to separate positive and negative ions from salt water. Ion separation was seen to noticeably improve as the strength of applied electric filed increased [84]. A combination of simulation strategy, DFT, MD, and MMC, was used to show that carbon graphite (111), as a heavy metal sorbent, has an efficient capability of removing silver (Ag), mercury (Hg), cadmium (Cd), palladium (Pd), and zinc (Zn). The results showed that the adsorption process is spontaneous and exothermic in nature. The maximum adsorption can be achieved in neutral to low acid medium [85]. The single layer C_2N nanosheet is more effective at purifying water because of its high porosity compared to other carbon-based membranes of a similar thickness and the presence of nitrogen atoms at the edges of each pore. An external voltage should be applied across the membrane to separate heavy metal ions through the nanosheet. According to the findings of radial pair distribution function analysis, the hydration shell of the cations like Cu^{2+} is responsible for the energy barrier that prevents ions from passing [86].

2.2 Boron nitride materials (BNM)

A unique hierarchical boron nitride structure that resembles an urchin created initially using a sample two-step process that involves the production of a comparable "core-shell" structured boron-containing precursor and heat catalytic chemical vapor deposition. This assembly is made up of free-rowing boron nitride nanotubes and crapy boron nitride nanosheets (**Figures 4** and **5**). To regulate the creation of BN hierarchical structure, a mixed growth mechanism of vapor–liquid–solid and vapor–solid is proposed. The distinctive structure demonstrates excellent Pb^{2+} and Cu^{2+} removal abilities in aqueous solutions. The superior adsorption ability of product is primarily due to the significant lattice defect, high-density edge active sites, extended interplanar space, and distinctive structural features. They contribute to structural stability and provide space for the heavy metal ions that have been absorbed [87].

The process of making hierarchical porous boron nitride nanosheets (hp-BNNSs) involved pyrolyzing a solution of boron acid and urea at 1000°C for 5 h. With pores that ranged in size from 1.1 nm to 40 nm and a larger specific surface area of 1145 m^2/g, the produced hp-BNNSs displayed a hierarchical porous structure. Heavy metal ions (Cu^{2+} and Ni^{2+}), industrial dyes (methylene blue and rhodamine B), and antibiotics (tetracycline) were all adsorbable on hp-BNNSs. This outstanding performance of the hp-BNNSs revealed their enormous potential for flexible water treatment [88].

Nanosheet-structured boron nitride spheres (NSBNSs) is created from solid B powders using a catalytic thermal evaporation process. The ultrathin nanosheets that make up NSBNSs are oriented radially, with the sheet edges toward the surface. The NSBNSs have a flexible capacity for adsorption and excel in removing heavy metal ions from water, oil, and dyes. The NSBNSs have adsorption capabilities for Cu^{2+} and Pb^{2+} that are equal to or greater than those of the adsorbents previously described [89].

A low temperature synthesis technique was employed to create new, few-layered boron nitride (BN-550) nanosheets, which were then used to quickly and effectively adsorb lead ions (Pb^{2+}). Through experiments regeneration and pH, easy recyclability and stability of the BN-550 were demonstrated. According to interference assays,

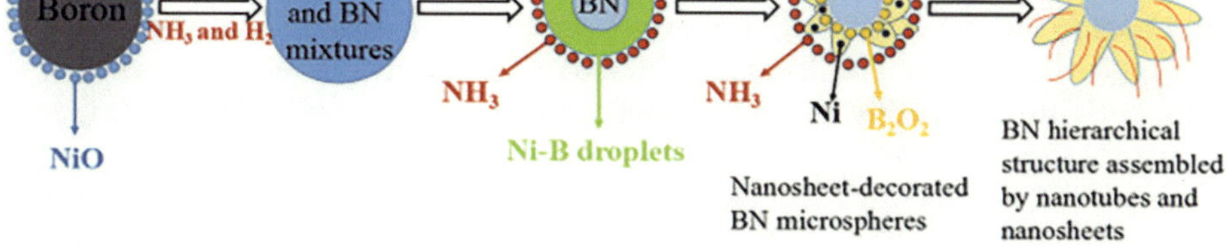

Figure 4.
An illustration of the BN hierarchical development process of structure using nanotubes and nanosheets [87].

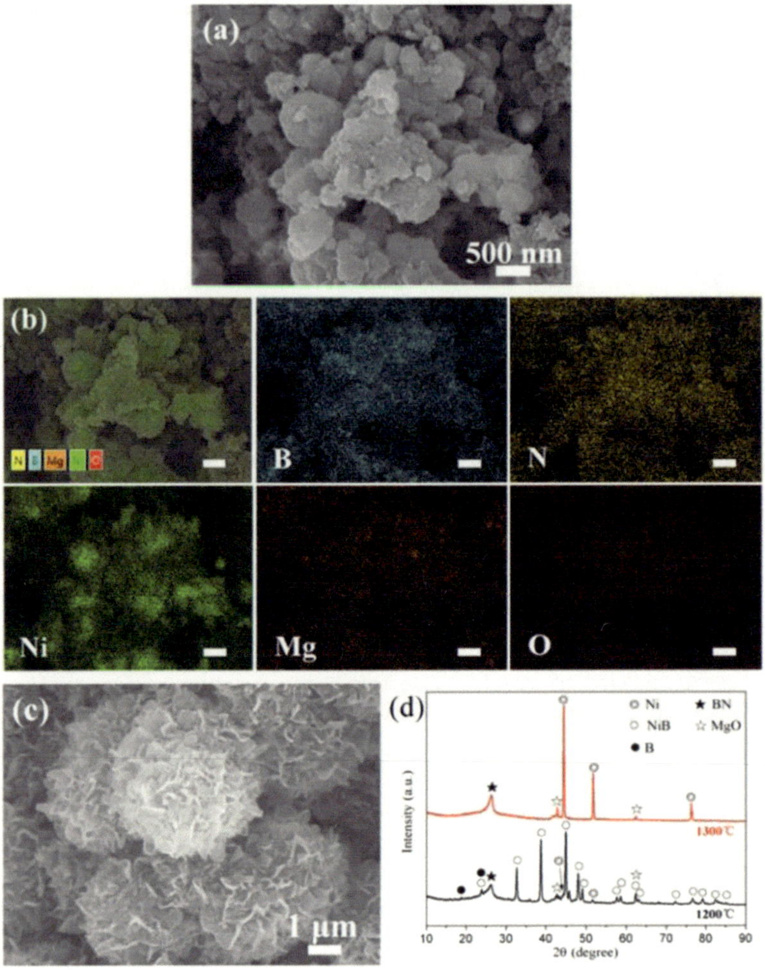

Figure 5.
(a) FESEM image of the primary product obtained at 1200°C and (b) corresponding EDS mapping images (scale bars represent 500 nm) (c) FESEM image of the primary product obtained at 1300°C (d) XRD patterns of the primary product obtained at 1200 and 1300°C [87].

the adsorbents had a significant affinity for Pb^{2+} even when other heavy metal ions including Ni^{2+}, Cu^{2+}, and Cd^{2+} ions interfered [90]. A straightforward two-step approach was used to easily create activated oxygen-rich porous boron nitride nanosheets (OBNNSs), which were rich in B-O bonds and boron atom vacancies. Unlike traditional doping and activation techniques, the oxygen dopants and boron atom vacancies in OBNNSs are mostly injected in situ and chemically activated during the production process. Due to their distinct polarity of B-O bonds and boron atom vacancies, OBNNSs have a high capacity and rate for absorbing metallic ions, outperforming both bulk and activated BN as well as many other absorbents. Aside from that, the extraordinary anti-oxidation, corrosion resistance, and structural stability of the employed OBNNSs make it simple to regenerate them using acidic elution [91].

Two researches were done using pyrolyzing a solution of melamine and boric acid to create boron carbon nitride nanosheets (BCN NSs) [82] and poly(ethyleneimine)-modified h-BNNSs (PEI-h-BNNSs) [83]. A number of products were produced by adjusting the molar ratio and heating temperature in order to improve the synthesis conditions, and 3:3 molar ratio at 550°C temperature was selected for the best Hg^{2+} and Pb^{2+} removal efficiency. BCN NSs have the highest possible adsorption capacity and demonstrated good adsorption performance toward Hg^{2+} and Pb^{2+} thanks to the effectively introduction of abundant function group, increased specific surface area, hydrophilicity, and electrostatic attraction ability. Additionally, BCN NSs exhibit exceptional chemical stability, and even after six cycles of adsorption/desorption, the adsorption capabilities were still greater than 90% [92]. Hexagonal boron nitride nanosheets (h-BNNSs) based on magnetic hybrid aerogels (MHAs) as a lightweight adsorbent were fabricated for Cr(VI), As(V), methylene blue (MB), and acid orange (AO) removal from aqueous solutions. Magnetic nanoparticles (Fe_3O_4 NPs) could be decorated on PEI-h-BNNSs because of PEI-h-BNNSs formation. Magnetic hybrid aerogel with large porous structures were generated by lyophilization treatment of PEI-h-BNNSs@Fe_3O_4 NPs- loaded PVA hydrogels, which has diverse and numerous functional groups, good super-paramagnetic, and a zero net surface charge. Due to these characteristics, this absorbent could be utilized to efficiently remove Cr(VI), As(V), MB, and AO with highest adsorption capacity [93].

Application of functionalized boron nitride nanosheets (BNNS-Fe_3O_4 nanocomposite) by Fe_3O_4 for As(III) and As(V) removal from contaminated water was studied. The nanocomposite could be easily separated from the solution under an external magnetic field because of its supermagnetic characteristic at ambient temperature. **Figure 6** shows that the highest adsorption capacity of the nanocomposite for As(III) and As(V) ions is determined to be 4 and 5 times more than that of the unmodified BNNSs. DFT simulations and experiments demonstrate that $As(OH)_3$ and $As(OH)_5$ bind to the BNNS-Fe_3O_4 nanocomposite more effectively than the raw BNNSs, which it showed higher adsorption capacity of the nanocomposite [94, 95].

Molecular dynamics simulation of hexagonal boron nitride (h-BN) showed that rationally designed h-BN membranes have excellent permeability, selectivity, and controllability for water desalination. Two key important features play an important role in controlling the water flow and ion rejection, which are size and chemistry of the pores. Nitrogen on the edges of pores were allowed higher flows than boron-lined pores. Particularly, two-dimensional h-BN with medium sized N4 pores exhibit exceptional water permeability and completely ion rejection that is several orders of magnitude higher than of standard membranes. Additionally, **Figure 7** shows that mechanical strain affects monolayer h-BN desalination performance with relatively small N3 pores, indicating that tensile strain can perfectly control desalination process [96].

Boron nitride nanosheet with functionalized pore obtained by passivating the atoms at the edges with fluorine and hydrogen atoms were simulated with molecular dynamics method [97, 98]. Cu^{2+}, Hg^{2+}, Pb^{2+}, and Cd^{2+} cations pass selectively through the functionalize pore of the boron nitride nanosheet because of an external voltage which was applied along the simulation. The results showed that heavy metal ions met an energy barrier and could not pass through the pores, but external voltage causes them to overcome the energy barriers and crossed the functionalized pores.

The adsorption capability of boron nitride and graphene nanosheets for arsenic [99, 100], Ni^{2+} [83], and Zn^{2+} [82] removal was done using molecular dynamics. Results for arsenic adsorption simulation show that the h-BN nanosheet has a strong interaction with arsenic. However, desorption of arsenic on the h-BN nanosheet displayed higher energies barriers than that of graphene. As a result, when compared to graphite nanosheets, the residence time of arsenic is approximately three times

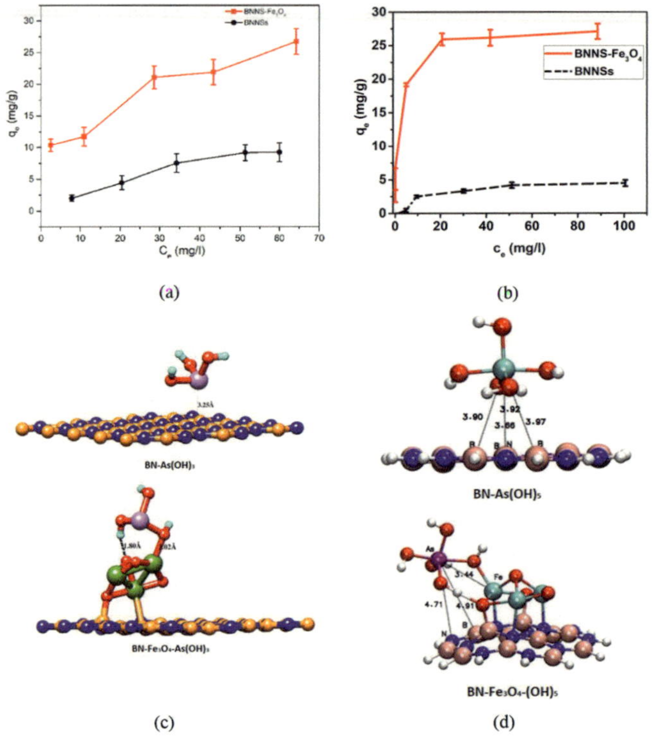

Figure 6.
Adsorption isotherm of (a) BNNSs (b) BNNS-Fe_3O_4 nanocomposite. Structures obtained from DFT optimization: (c) BNNSs-As $(OH)_3$ and BNNS-Fe_3O_4-As $(OH)_3$ nanocomposites (d) BNNSs-As $(OH)_5$ and BNNS-Fe_3O_4-As $(OH)_5$ nanocomposites (a, c): https://pubs.acs.org/doi/10.1021/acsomega.9b04295. Note: Further permissions related to the material excerpted should be directed to the ACS) [94, 95].

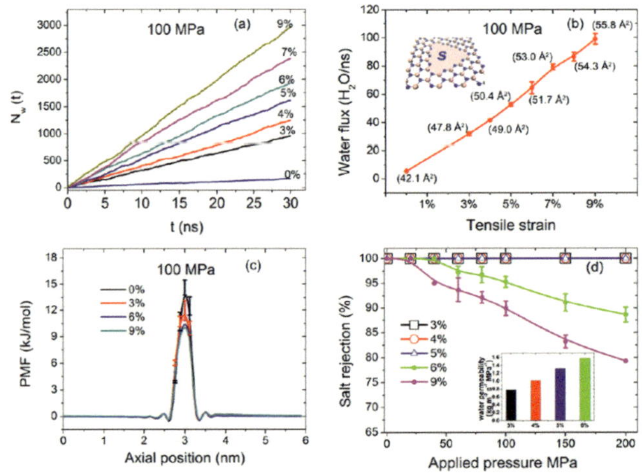

Figure 7.
(a) The number of water molecules passing through the strained N3 membrane at 100 MPa. (b) Water flux vs. strain at 100 MPa. The inset shows the enlarged pore, with a shadow indicating the pore surface area (S), and the values of S are provided in parentheses when tensile strain is applied. (c) PMF of water across the N3 membrane without strain (0%) and with a strain of 3, 6, and 9%. (d) Salt rejection vs. pressure by the N3 membrane under different tensile strains, and the inset plots show the water permeability under 3 ~ 6% strain [96].

higher on h-BN nanosheet in comparison with graphene. Also, we should note that the adsorption behavior of ions is influenced by presence of limited charges on B and N atoms in the h-BN [99]. However, for simultaneous adsorption of methylene blue and arsenic, graphene indicated better efficiency for wastewater treatment compared to hexagonal boron nitride and hexagonal boron carbon nitride [100]. To remove Ni^{2+} and Zn^{2+} from aqueous solution, molecular dynamic simulation was done with graphene and boron nitride with the functionalized pores under an external voltage. Studies showed that these functionalized nanosheet membranes have great functionality for heavy metal ions removal [82, 83].

2.3 Transition metal dichalcogenides (TMDs)

Molybdenum disulfide (MoS_2) has recently noted as promising heavy metal removal from aqueous solution nano-based adsorbent, which has the potential to provide an alternative to conventional water decontamination technologies. The trade-off between mercuric removal capacity and overall MoS_2 adsorbent stability driven by MoS_2 generation parameters was studied. Flower-like MoS_2 films onto planar alumina support were grown with a bottom-up hydrothermal synthesis setup at different growth temperature. The results showed that low growth temperature could generate amorphous supported MoS_2 enhancing mercuric removal, but lowering the chemical stability, which has considerable byproduct molybdate leaching [101].

A novel removal mechanism was clarified that includes a reduction–oxidation (redox) reaction between heavy metal ions and ce-MoS_2. Ag^+ ions were used as model species, and to demonstrate this novel removal mechanism, after removal experiments, the Ag contained in MoS_2 was separated using centrifugation and then characterized. The Ag peaks in XRD patterns, indicating a redox reaction between Ag^+ and ce-MoS_2. Ag^+ is a mild oxidant and could oxidize the MoS_2 nanosheets to soluble molybdate and sulfate/sulfite ions with reduce itself to metallic Ag(0) particle. Because the key oxidation products are liquid, non-toxic molybdate, and sulfur species, which do not shape an insulating oxide layer to passivate the membrane surface, the surface is still active to effectively recover metallic particles. Hence, precious heavy metals can be effectively removed from wastewater and recovered using MoS_2 nanosheets membrane [102].

An ultrathin MoS_2 nanosheets (MoS_2 NS) were created with combination of quenching process and liquid-based exfoliation. The resulted MoS_2 Ns retained hexagonal phase (2H-MoS_2) and demonstrated thin layer structure with hidden wrinkle. The adsorption tests showed that this ultrathin MoS_2 NS revealed significant adsorption on heavy metals and dyes. The main adsorption mechanism implicated of physical hole-filling effects and electrostatic interactions [103]. Additionally, the MoS_2 NS can be combined with a facile one-step hydrothermal method without addition of templates or surfactants. Enlarged interlayer spacing and multiple defects on the basal planes characteristics of the products proved to be crucial in the removal of Cr(VI). Adsorption and reduction are the synergism of removal process, which the MoS_2 NS can adsorb Cr(VI) and reduce some fraction of adsorbed Cr(VI) to low toxic Cr(III) simultaneously. This process not only remove the Cr(VI) from aqueous solution but also diminished the toxicity of Cr(VI) [104].

MoS_2 NS has the capability of mercury removal, but successful attempt to apply MoS_2 NS for mercury removal are completely few, because the most of sulfur atoms are located inside the bulk of MoS_2 and therefore inaccessible for mercury ions. To resolve this problem, MoS_2 NS with widened interlayer spacing are able of mercury removal [105].

MFe_2O_4 (M = Mn, Co)-MoS_2-carbon dots (CDs) nanohybrid composites (MnFMC and CoFMC) [106] and millimeter-sized nanocomposite MoS_2–001 [107] were used to Pb^{2+} removal. Exploit of high surface available of MoS_2 for enhancing sequestration of target metal pollutants and cation substitution of abundant functional groups on the surface of CDs and MFe_2O_4 nanoparticles, the product composites demonstrate high adsorption performance and preferential Pb^{2+} sorption with high concentration of competing cations (Ca^{2+}/Mg^{2+}) [106]. MoS_2 nanosheet was loaded into a polystyrene cation exchanger D-001 by a facial hydrothermal method. The as-prepared nanocomposite demonstrated extraordinary adsorption capacity and high selectivity for Pb^{2+} sorption. In addition, NaCl-$EDTANa_2$ solution can regenerate this nanocomposite without any loss in adsorption capacity [107]. Besides the experiments, which is showing the strong selectivity/affinity of MoS_2 nanosheets toward Pb^{2+}, a DFT simulation was conducted to study the adsorption mechanism (**Figure 8a**). The results demonstrated that the cation selectivity of MoS_2 is in sequence of $Pb^{2+} > Cu^{2+} > > Cd^{2+} > Zn^{2+}$, $Ni^{2+} > Mg^{2+}$, K^+, Ca^{2+} (**Figure 8b**). The membrane made of layer-stacked MoS_2 nanosheets had a high water flux, but it is also effectively reduced the concentration of Pb^{2+} in drinking water [108].

Metallic phase (1 T) and semiconducting phase (2H) of MoS_2 nanosheet were studied experimentally and theoretically for uptake of Pb^{2+}, Cu^{2+}, and Hg0 ions. Computational results showed that 1 T-MoS_2 could remove Pb^{2+} and Cu^{2+} more preferably that 2H-MoS_2. Facile hydrothermal reaction is the synthesis mechanism of both 1 T and 2H MoS_2 nanosheets. In addition, 1 T-MoS_2 has demonstrated superior properties over 2H-MoS_2, such as ultrafast adsorption kinetics and strong anti-interference activity toward other existing cations [109]. Also, mechanism of 1 T-MoS_2 in Hg0 uptake was studied by DFT simulation, and the results indicated that the adsorption between Hg0 atoms and the nanosheet is chemisorption mechanism and the strongest adsorption configuration is on the top of the Mo atom position [110].

Theoretical and experimental showed that MoS_2 surface could be more reactive in the presence of S-vacancy defects, and this leads to strong adsorption energy of H_2o, Hg^{2+} and O_2 on MoS_2 surface. In addition, surface oxidation can occur when there are enough reaction sites for the Hg^{2+} adsorption and surface oxidation simultaneously [111]. MoS_2 nanosheet with various pores were investigated to evaluate their selectivity to heavy metal ions (**Figure 9a,c**). When the pore size of membrane matched

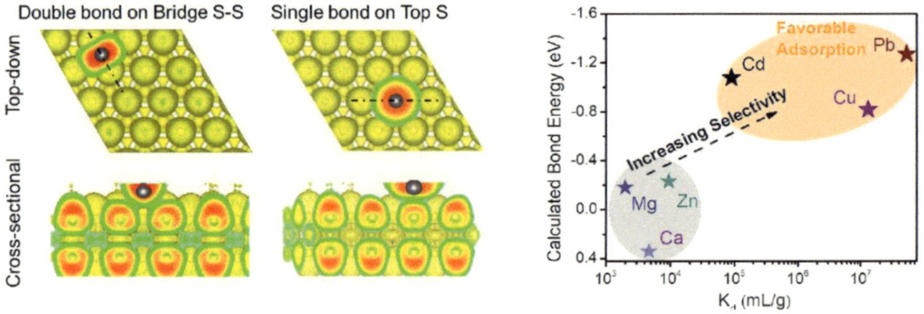

(a) (b)

Figure 8.
Investigation of Pb^{2+} adsorption. a) DFT simulation b) the maximum bond energy for the binding between cations and MoS_2 [108].

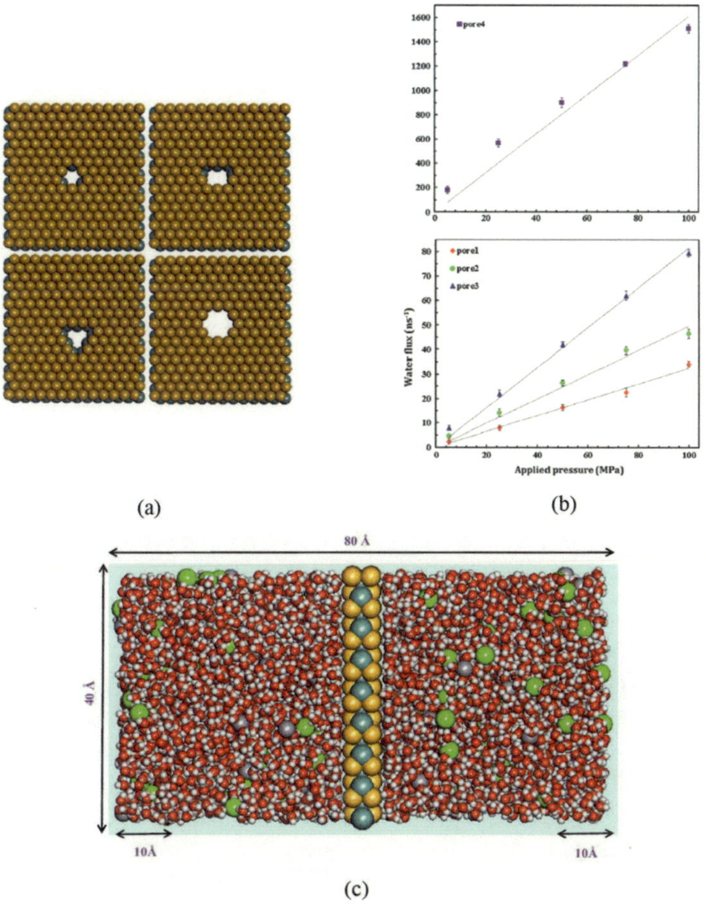

Figure 9.
Studying the adsorption capability of MoS_2 with different pores. (a) Different pores of MoS_2 (b) the water flux in various applied pressure for designed pores (c) schematic representation of the simulation framework [112].

that of ion hydration shell, the ions were rejected. **Figure 9b** showed that because of fish-bone structure of MoS_2, the molybdenum pore (pore3) had a high adsorption, and water molecules could pass through it under external pressure. According to simulations, Hg^{2+} ions encountered a significant energy barrier and were unable to pass through created pores. The finding indicted that the type and size of pores affected the ability of reject mercury ions of MoS_2 nanosheet [112].

2.4 Layered double hydroxides (LDHs)

A nanocomposite of graphene/MgAl-layered double hydroxides (G-MgAl-LDH) was created by hydrothermally reacting $Al(NO_3)_3.9H_2O$, $Mg(NO_3)_2.6H_2O$, graphene oxide (GO), and urea. In addition to efficiently reducing GO, urea can also cause the formation and in situ development of LDH crystallites on graphene nanosheets. In the nanocomposite, the LDH and graphene layers were both exfoliated. By heating G-MgAl-LDH to a higher temperature, it was simple to make the calcined G-MgAl-LDH, which could be employed as a nanoadsorbent to remove hexavalent chromium from an aqueous solution. As eliminating Cr(VI) from aqueous solution, calcined

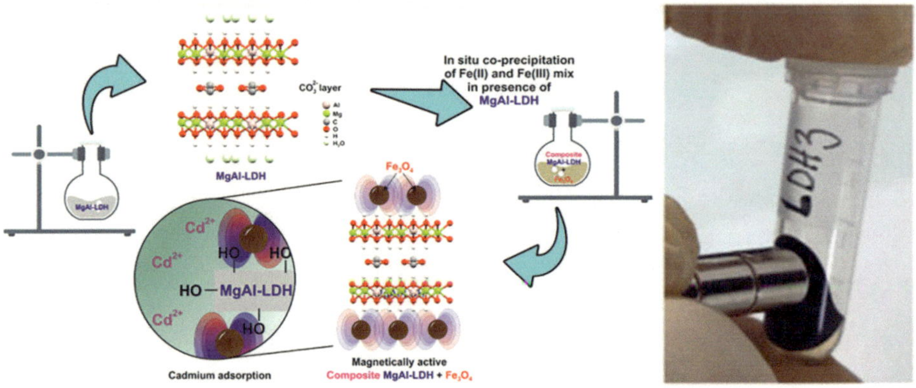

(a) (b)

Figure 10.
(a) Mechanism of adsorption for composite MgAl-LDH + Fe_3O_4 (b) the attraction effect of the magnetic field on magnetic MgAl-LDH + Fe_3O_4 composite. The magnetic fields are generated by NdFeB magnets [114].

G-MgAl-LDH shown greater adsorption effectiveness with a lower dose when compared to virgin Mg-Al-LDH. The findings of the adsorption kinetics, isotherm, and thermodynamics showed that the surface adsorption of graphene, the memory effect of calcined LDH, and the synergistic contributions from each component were engaged in the adsorption process of Cr(VI) [113]. A study demonstrates the topological change of 3D hierarchical MgAl-layered double hydroxides in the presence of Fe_3O_4 nanoparticles to produce highly distributed magnetic hetero-nanosheets. The 10 nm Fe_3O_4 magnetic nanoparticles were disseminated throughout the surface of the MgAl-LDH lamella to create the platelet-like hierarchical nano-heterostructures. The biphasic self-assembled nanocomposite was effectively employed as an adsorbent for the ambient temperature decontamination of wastewater spiked with cadmium. The synthesized materials are good candidates for adsorbents in Cd(II) adsorption procedures with the highest adsorption capabilities, according to thorough evaluation (**Figure 10**). The regeneration research also showed how well the adsorbents performed following three cycles of adsorption/desorption [114].

By self-assembling between oppositely charged 2D nanosheets of LDH and –MnO_2 nanosheets in the presence of zwitterionic histidine biomolecules, layer-by-layer LDH/histidine/MnO_2 (LDH/his/MnO_2) nanohybrids were created. Exfoliation of highly crystalline MgAl-LDH and bulk MnO_2, respectively, resulted in the preparation of single-layer LDH and –MnO_2 nanosheets. The massive capacities for Pb(II) and As(V) and extremely high distribution coefficients (Kd) for Pb(II) and As(V) amply show the effectiveness of the suggested method for layer-by-layer synthesis of LDH/His/MnO_2 material. The results clearly demonstrate that layer-by-layer building of organic molecules and inorganic 2D nanosheets improved the simultaneous removal of the heavy metal ions Pb(II) and oxyanion of As(V) [115]. In a new study, MgAl-layered double hydroxide nanosheets were created using a straightforward hydrothermal method and assembled on graphene oxide and its magnetic product (Fe_3O_4@GO). The adsorbents combined diverse functional groups with sizable distinct surface area, according to characterization results. Under neutral and weakly alkaline conditions, GO and Fe_3O_4@Go successfully removed heavy metals ions. The primary mechanisms are the invertible substitution of divalent metal cations, metal hydroxides and metal carbonates precipitation reactions, and surface complexation of hydroxyl groups. So, new knowledge about heavy metal environmental remediation is provided by the improved adsorption capacity and mechanistic study [116].

A layered double hydroxide (MgAl-LDH) nanosheets that could effectively and simultaneously capture heavy metal cations and oxyanions from wastewater were created using pinewood sawdust-derived engineered biochar. The adsorption performance of loaded MgAl-LDH and biochar toward cationic and anionic contaminants, such as Pb^{2+} and CrO_4^{2-}, is significantly improved by the synergetic effect. Complexations with surface functional groups were primarily responsible for the elimination of Pb^{2+}. While CrO_4^{2-} can be transformed into Cr^{3+} by functional groups for removal of oxyanions, the resulting Cr^{3+}, which could then replace Al^{3+} through morphic substitution, formed an MgCr-LDH structure. The prepared nanocomposites were found to be suitable for the maximum adsorption capacity of Pb^{2+} ions from an inaqueous solution due to their distinctive 3D porous network and abundance of oxygen containing functional groups. After 180 minutes of desorption experiments at pH 4.5, equilibrium was attained. Desorption experiments made it possible to regenerate adsorbents for use in next time and to recover copper ions [117].

An ultrathin MgFe-LDH nanosheets provide excellent Cu^{2+} ion mineralization in aqueous solutions and soils. Surface adsorption and isomorphous substitution are used in the removal of Cu^{2+} to transform MgFe-LDH into Cu(Mg). By adsorption, the Cu(Mg) Fe-LDH product can either be used directly to get rid of phosphate anions and zero dyes, or it can be converted into metallic copper through leaching-electroreduction. Even in acidic soils, it has been demonstrated that the release of Mg^{2+} ions by MgFe-LDH during Cu^{2+} removal promotes crop growth by acting as magnesium fertilizer [118].

Nanosheets conforming of two-dimensional nanomaterials made up of Ca^{2+}, and Y^{3+} cations and carbonate(CO_3^{2-}) anions in the interlayer with a invariant consistence and lengths of around 10 μ m have been successfully synthesized in a hydrotalcite subcaste structure, known as a layered double hydroxide, using a facile hydrothermal system. The as-prepared CaY-CO_3^{2-} layered double hydroxide materials demonstrate outstanding affinity and selectivity for metal ions including Cr^{3+}, Ni^{2+}, Cu^{2+}, Zn^{2+}, Pb^{2+}, Cd^{2+}, and Hg^{2+} as well as metalloid As^{3+}. The adsorption of all of the heavy metal ions from the aqueous solution was set up to be exceptionally rapid and highly selective, with more than 95% removal achieved within 30 min. The as-prepared adsorbent has excellent chemical stability, which is included they retain their well-defined lamellar shapes indeed under mildly acidic conditions [119].

Thio-functionalized adsorbents are different from conventional ones that are based on are oxygen group. In order to create layered double hydroxide (LDH) as Mn-MoS_4, MoS_4^{2-} anions were inserted into lamellar layers of LDH. This adsorbent was used to remove heavy metals because it was highly stable, effective, and selective. Mn-MoS_4 was ranked at the most effective adsorbent for the removal of such metals so far due to its record high distribution coefficient, fast kinetics, and enormous saturated uptake capacities for Hg^{2+}, Ag^+, and Pb^{2+}. Additionally, Mn-MoS_4 is able to effectively control the concentrations of Hg^{2+}, Ag^+ ions from tap water, lakes, and industrial wastewater far below the limits for drinking water. It can also completely withstand the effects of extremely high background electrolytes (Ca^{2+}, Mg^{2+}, Na^+, Cl^-, NO^{3-}, and SO_4^{2-}). The dominant chemical adsorption mechanism, which coordinates the inner spheres of thio group as soft Lewis acids and Hg^{2+}, Ag^+, and Pb^{2+}, is where these remarkable characteristics come from. More intriguingly, intercalated MoS_4 is protected by LDH levels. Mn-MoS_4 is made easier to store and use than other adsorbents owing to the ability to protect MoS_4^{2-} anions from oxidation [120]. A hierarchical porous hybrid monolith, also known as NiFe-MoS_4^{2-}-LDH/CF, is created by homogenously immobilizing double hydroxide (LDH) nanosheets on a carbon foam (CF) substrate and then intercaling MoS_4^{2-} ions into the interlayers. The developed NiFe-MoS_4^{2-}-LDH/CF hybrid monolith, which has ultrahigh sorption capacities, is found to be highly effective for sequestering Hg^{2+}, Pb^{2+}, and Cu^{2+} due to its abundance of binding sites, strong affinity and excellent pore accessibility. As evidenced by more than 99% removal rates within 5 min, the metal ions uptake kinetics of this nanoadsorbent are extremely high. More importantly, in the presence of different interfering ions, the developed adsorbent exhibits exceptional sensitivity for the target heavy metals with high distribution coefficients [121].

Due to poor adsorbent-metal interaction, typical sorption technologies are ineffective. In order to achieve this, a simple complexation technique was used to extract dangerous metals from an aqueous solution to create layered cationic platform material loaded with phosphonate. Sample experiments, detection methods, and simulation calculation were used to examine the interactions between clay and water in order to learn the removal mechanisms. The functionalized layered double hydroxide, in particular, had excellent chelation adsorption properties with Zn^{2+} and Fe^{3+}, and model fitting showed that chemisorption and monolayer interaction were responsible for the process. The interaction region indicator, a novel concept, was used to describe poor interaction and coordinate bonds after the molecular dynamics simulation (**Figure 11**) of the interfacial interaction between the phosphonate and clay surfaces was evaluated. The orbital interaction diagram graphically conveyed the in-depth

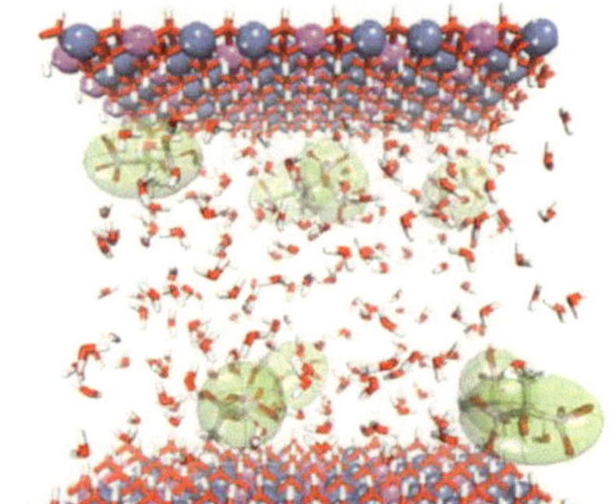

(a)

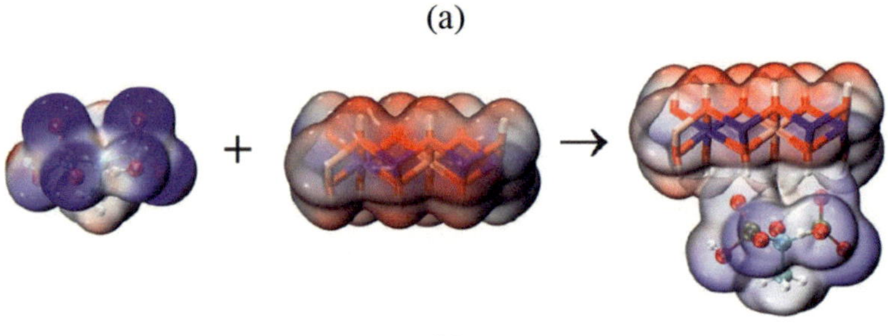

(b)

Figure 11.
MD simulation of adsorbent of functionalized layered double hydroxide (a) final configuration of MD model (b) penetration analysis of the ESP on the vdW surface [122].

understanding of the chelation mechanism. Severe toxicity analysis, the adsorption column test, and the regeneration of the spent adsorbent all showed that the synthesized material has enormous potential for use in real-world toxic pollution treatment [122]. Phosphonates are environmental-friendly materials and can be possible employed to improve the loss efficiency of the clay materials. In an experimental and DFT study, Zn-Al-layered double hydroxide (LDH) intercalated with amino trimethylene phosphonic acid (ATMP) by a facile technique and employed as an adsorbent for Cu^{2+} and Pb^{2+} from wastewater. The weak interaction investigation indicated that interactions between ATMP and LDH are primarily dominated by H-bond. These benefits recommended that modified LDH can be a tempting adsorbent for adsorption of harmful metal ions [123].

2.5 MXenes

Application of two-dimensional metal carbides and nitrides (MXenes) in water/wastewater treatment has been receiving more attention. Theoretical and experimental studies were done on mercury and copper ions with MXene nanosheets [124–127]. The functionalization of MXenes to growth their balance at the same time as demonstrating excessive pollutant elimination can facilitate sustainable water/wastewater treatment processes. A quite solid magnetic titanium carbide (Ti_3C_2Tx) MXene nanocomposite (**Figure 12a**) (MGMX nanocomposite) changed into correctly synthesized via a facile hydrothermal method and changed into examined for aqueous-phase adsorptive elimination of mercuric ions. Great Hg(II) adsorption in a high concentration of pH conditions and an outstanding experimental Hg(II) uptake capacity were exhibited with the MGMX nanocomposite. In the adsorption/desorption investigation, the MGMX nanocomposite showed reusable capacity for 5 cycles. The MGMX nanocomposite is an efficient sorbent for the removal toxic Hg(II) for water purification due to stability, hydrophilic nature, available adsorptive surfaces, and easy separation after reaction [124]. Using a simple technique, two-dimensional titanium carbide MXene core (Ti_3C_2Tx) shell aerogel spheres (MX-SA) for mercuric ion removal were created (**Figure 12b**). The synthesized microspheres make an excellent adsorbent for the removal of heavy metals from water due to their distinctive internal structures, deep porosities, big specific surface areas, oxidized functional groups of MXene nanosheets, and available active binding sites. The adsorbent has excellent single- and multi-component removal efficiencies, with Hg^{2+} being 100% efficient and five heavy metal ions being more than 90% efficient. Under extremely low pH conditions (0.5–1.0 M HNO_3), the synthesized materials are highly effective at removing Hg^{2+} and have excellent reproducible qualities. This adsorbent can also be used in column-packed devices due to its small size and spherical shape [125].

Multilayered oxygen-functionalized Ti_3C_2 (also known as M-Ti_3Ox) nanosheets were ready to uptake Hg(II) from water. The M-Ti_3C_2 has shown to be extremely fast adsorption kinetics, impressively high capacity, high selectivity, and a wide working pH range (3–12) characteristics. This exceptional Hg(II) removal is explained by the distinct interaction (e.g., adsorption coupled with catalytic reduction), according to density functional theory calculations and experimental characterizations. In particular, Ti atoms on {001} facets of M-Ti_3C_2 prefer to adsorb Hg(II) in the form of HgClOH, which then undergoes hemolytic cleavage to form radical species (e.g., OH and HgCl). On the edges of M-Ti_3C_2, the HgCl radicals immediately dim the crystallize to form Hg_2Cl_2. Simple thermal treatment can effectively recover up to 95% of dimeric Hg_2Cl_2. Notably, M-Ti_3C_2 has oxidized to TiO_2/C nanocomposites as a result

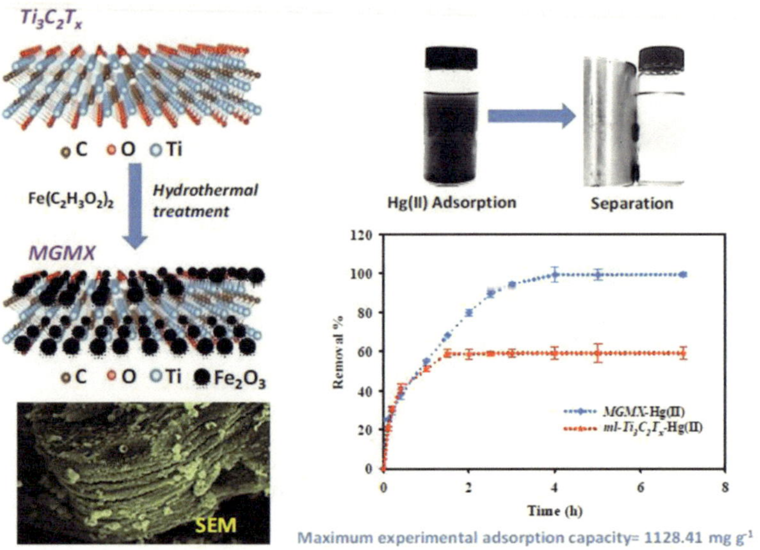

(a)

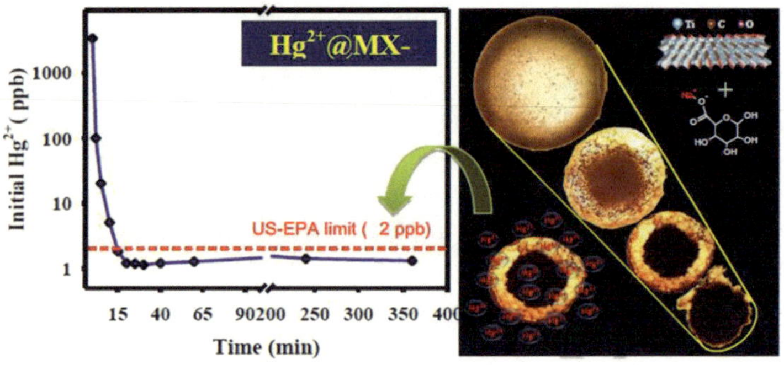

(b)

Figure 12.
Experimental studies on Ti_3C_2Tx for Hg(II) removal. a) Magnetic titanium carbide (Ti_3C_2Tx) MXene nanocomposite b) two-dimensional titanium carbide MXene core (Ti_3C_2Tx) shell aerogel spheres (MX-SA) [124, 125].

of the adsorbed OH and energy released during the distinct interaction. Additionally, TiO_2/C nanocomposites have demonstrated superior performance to Degussa P25 in terms of photocatalytic degradation of organic pollutants. M-Ti_3C_2 is a superb candidate for quick or urgent Hg(II) removal and recovery due to these exceptional qualities and its mercuric recyclable nature [126].

Due to their large specific surface area, hydrophilicity, and distinctive surface functional properties, delaminated (DL)-$Ti_3C_2T_x$ displayed excellent Cu removal ability. Reductive adsorption of Cu^{2+}, which resulted in the formation of Cu_2O and CuO species, was facilitated by oxygenated moieties in MXene layer structure. When compared to multilayer (ML)-$Ti_3C_2T_x$, DL showed a higher and quicker Cu uptake. The endothermic nature of the adsorption process was discovered by thermal analysis. Adsorption capacity of DL-$Ti_3C_2T_x$ was 2.7 times greater than an activated carbon

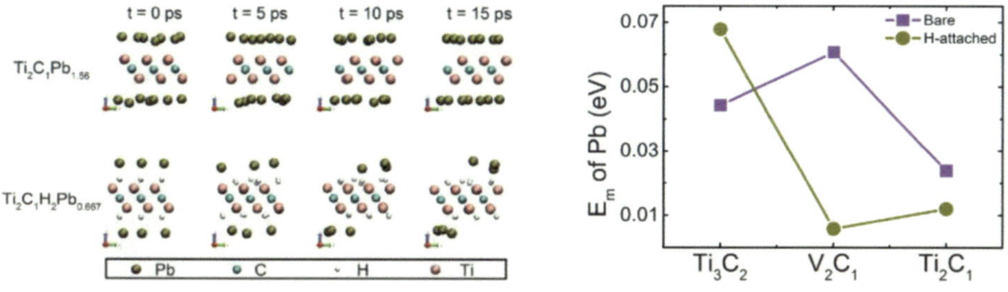

(a) (b)

Figure 13.
DFT simulation of different two-dimensional materials MXenes, such as Ti_3C_2, V_2C_1, and Ti_1C_1. (a) Schematic of fully Pb adsorbed Ti_2C_1 and $Ti_2C_1H_2$ that endure 15 ps ab-init MD simulation at 300 K. (b) Comparison of migration energy barrier Em of a single Pb atom on bare and H-decorated samples [128].

Nanosheet	Advantages	Limitations	Ref.
Carbon products	1.sp2 domains are re-established 2.Electron transport property was improved 3.Colloidal stability 4.Higher dispersibility in water 5.The abundance of oxygenated groups	1.Low colloidal stability 2.Oxygen-containing functional groups are found to be of low density 3. Limited sorption sites	[129]
BNM	1.Large surface area 2.Polarity 3.Hydrophobic 4.Excellent corrosion and oxidation resistance 5.Very high permeability 6.Stable under thermal and chemical shocks	1.Poor dispensability 2.Low chemical reactivity	[130–133]
TMDs	1.Large number of adsorption sites 2.Low water affinity 3.Photocatalytic properties 4.High surface area 5.Multilayer filtration 6.Easy to functionalize 7.High antibacterial activity with no toxicity 8.Medium permeability 9.Stable in acidic and alkaline solutions	1.Ineffective control over interlayer spacing 2.Hydration of membrane is required at all times for efficient water transport 3.Functionalization is required to achieve at least 50% salt rejection	[134–136]
LDH	1.Selective absorbability 2.Liquid–solid separation 3.Stability 4.Recyclability	1.The lack of functional groups and structural components of pristine LDH 2. Low surface area 3. Small particle size	[137, 138]
MXene	1.Surface tenability 2.Electrical conductive properties 3.Hydrophilicity of MXene flakes 4.High permeability 5.Stable under thermal and chemical shocks	1.The inferior durability 2.lesser-biocompatibility 3.High tendency for aggregation 4.Improper reusability 5.Ineffective control over interlayer spacing	[139]

Table 1.
Advantages/disadvantages of nanosheets using in heavy metal removal.

that was readily available on the market. These results show how tempting 2D MXene nanosheet are for removing hazardous metals from water [127].

The adsorption characteristics of Pb and Cu on recently synthesized two-dimensional materials MXenes, such as Ti_3C_2, V_2C_1, and Ti_1C_1, are studied using density functional theory calculation (**Figure 13a**). The majority of the studied MXenes have excellent capability to adsorb Pb and Cu, particularly Ti_2C_1. The practical groups attached to surface of MXenes are sensitive to both the adsorption energies and capacities. Ti_2C_1 remains stable at ambient temperature after adsorbing Pb atoms (**Figure 13b**), according to ab initio molecular dynamics (ab-initio MD) simulations. These results indicated that these newly emerging two-dimensional MXenes are promising candidates for wastewater treatment and ion separation [128].

Not only nanosheets have advantages in heavy metal removal compared to other adsorbents, but also there are some disadvantages for each kind of nanosheets. **Table 1** illustrates advantages and disadvantages of different nanosheets described in this chapter for heavy metal removals.

3. Adsorption mechanism

Adsorption method is widely used for the removal of heavy metals from water and wastewater. Adsorption has some benefits over traditional techniques, such as reducing chemical and biological waste, being less expensive, being more effective, being able to regenerate the adsorbents, and possibly recovering metals [140, 141]. There are three main mechanisms for sorption of heavy metals and are physical sorption, ion exchange, and electrical interactions [142]. Study of adsorption isotherms and kinetics could introduce the sorption mechanism of heavy metals using nanosheet. Adsorption isotherm shows the adsorption amount of adsorbate using equilibrium concentration. Freundlich and Langmuir isotherms are the most famous isotherms, which are used to investigate the adsorption equilibrium data [143].

$$\frac{C_e}{q_e} = \frac{C_e}{q_m} + \frac{1}{K_L q_m} \tag{1}$$

$$\log q_e = \log K_f + \frac{1}{n}\log C_e \tag{2}$$

Eqs. (1) and (2) are the linear equation of Langmuir and Freundlich isotherms. In these equations C_e (mg/L) is the concentration of heavy metals ion after equilibrium, q_e (mg/g) is the adsorption capacity of the adsorbents at equilibrium, q_m (mg/g) is the maximum adsorption capacity, K_L (L/mg) is the Langmuir constant, K_F is the Freundlich constant related to the sorption capacity, and n is the Freundlich constant associated to the adsorption intensity. **Table 2** illustrates the result of adsorption study of different nanosheets for heavy metal removal.

Table 2 shows the correlation coefficient (R^2) for Langmuir and Freundlich isotherm methods. According to the results, the Langmuir model is well fitted for MgAl-LDHs, and it shows that the adsorption of heavy metals had occurred uniformly and following the monolayer adsorption instead of multilayer adsorption [146]. The R^2 value of the Langmuir model was higher than Freundlich model in the cases of Pb(II) and Cu(II) adsorption onto 1 T-MoS_2 and of Cu(II) adsorption onto 2H-MoS_2,

Nanosheet	Target heavy metal	Langmuir isotherm			Freundlich isotherm			Ref.
		q_m	K_L	R^2	K_F	n	R^2	
		mg.g^{-1}	L.mg^{-1}		mg.g^{-1}			
Few layered magnetic graphene oxide	Cu(II)	2667.89	0.0178	0.984	14.372	0.862	0.999	[144]
	Cd(II)	2416.86	0.0571	0.959	48.811	0.508	0.995	
Boron Nitride	Cd(II)	236	0.112	0.984	1.27	4.367	0.987	[145]
1 T-MoS_2	Pb(II)	129.92	2.475	0.974	60.03	6.094	0.897	[109]
	Cu(II)	73.75	3.115	0.991	38.19	7.727	0.882	
2H-MoS_2	Pb(II)	55.11	2.385	0.802	23.71	5.838	0.997	
	Cu(II)	41.80	0.0038	0.996	0.72	1.679	0.977	
MgAl-LDHs	Cr(VI)	63.8	38.808	0.999	51.800	11.062	0.753	[146]
	Ni(II)	92.3	1.874	0.999	64.096	8.440	0.796	
Two-Dimensional $Ti_3C_2T_x$ MXene	Cu(II)	86.53	0.034	0.900	20.29	4.05	0.962	[127]

Table 2.
Study of isotherm of heavy metal removal with nanosheets.

indicating the adsorption process was more likely to be the monolayer accompany with a fraction of multilayer chemical adsorption. However, when Pb(II) was adsorbed onto 2H-MoS_2, the higher R^2 value of the Freundlich model indicates the existence of heterogeneous surface adsorption and non-uniform active sites [109]. On the other hand, the correlation coefficient of Freundlich isotherm model for few layered magnetic graphene oxide, boron nitride, and two-dimensional MXene nanosheets is higher than Langmuir model. The Freundlich model assumes that the interactions at solid–liquid interfaces involve the homogenous adsorption of the adsorbate on the heterogeneous and multilayer surface of the adsorbent [127, 144, 145].

The adsorption procedure generally consists of four stages. A mass transfer from the outside takes place first (heavy metal ions from bulk solution to the surrounding layer of adsorbent). The external diffusion to the exterior surface occurs in step two, and the internal diffusion of the adsorbate to the internal adsorption location occurs in step three. Adsorption, which takes place on the surface of the adsorbents, is the last procedure. A novel assumption of the rate-determining phase leads to the development of a unique kinetic model. In order to understand the kinetics of heavy metal adsorption on adsorbents, pseudo-first order and pseudo-second order kinetic models are frequently used [143].

$$\log(q_e - q_t) = \log q_e - \frac{K_1}{2.303}t \tag{3}$$

$$\frac{t}{q_t} = \frac{1}{K_2 q_e^{\,2}} + \frac{t}{q_e} \tag{4}$$

Eqs. (3) to (4) are the pseudo-first order and pseudo-second order. q_t (mgg^{-1}) is the adsorption capacity at respected time, K_1 (1/min) and K_2 (gmg^{-1} min^{-1}) are rate constants of pseudo-first order, pseudo-second order models. **Table 2** shows the result of adsorption kinetics study of heavy metal removal using different nanosheets.

The pseudo-first order kinetic model implies that physisorption—in which sorption happens only through feeble Van der Waals forces and without the presence of chemical bonds—is the mechanism by which adsorption takes place. This form of adsorption is simpler to regenerate because the adsorption is also readily reversible. According to the pseudo-second order, two responses can happen one after the other or concurrently. The initial response occurs quickly and quickly achieves equilibrium. The second reaction, in contrast, proceeds slowly and takes longer to achieve balance. Pseudo-second order suggests that chemisorption is how the adsorption takes place. From pseudo-first order kinetics, it is inferred that the bonding results from electronic sharing and that the transfer between adsorbents and adsorbate is comparatively stronger than that of the physisorption way [143].

According to the correlation coefficients in **Table 3**, it can be concluded that the majority of the heavy metal adsorption on nanosheet adsorbents had fit the pseudo-second order kinetic model well. This suggests that the surfaces of these nanosheets are heterogeneous and that chemisorption is likely to be the rate-limiting step during the adsorption processes.

Typically, electrostatic interaction, hydrophobic interaction, chelation, ion exchange, hydrogen bonding, precipitation, reduction, complexation, interaction, or weak Van der Waals interaction are involved in the absorption of heavy metals on adsorbents. One or a mixture of two or more of these interactions drive the adsorption process. The pH of the solution, the adsorbent's textural characteristics, and the chemical structure of the target compounds are just a few of the variables that affect the process [147].

Table 4 shows the interaction mechanisms involved in the removal of heavy metals using various nanosheets as the adsorbents. Although there are many interaction

Nanosheet	Target heavy metal	Pseudo-first order			Pseudo-second order			Ref.
		K_1	q_e	R^2	K_2	q_e	R^2	
		min^{-1}	$mg.g^{-1}$		$mg^{-1}.g.min^{-1}$	$mg.g^{-1}$		
Few layered magnetic graphene oxide	Cu(II)	1.361	3345.63	0.878	0.0034	1146.91	1.000	[144]
	Cd(II)	1.119	30.47	0.766	0.0683	403.23	1.000	
Boron Nitride	Cd(II)	0.524	111.7	0948	0.0001	193.1	0.999	[145]
1 T-MoS_2	Pb(II)	1.243	55.20	0.997	0.088	55.68	0.999	[109]
	Cu(II)	0.364	50.66	0.935	0.011	53.42	0.995	
2H-MoS_2	Pb(II)	0.014	43.56	0.991	0.0003	49.85	0.998	
	Cu(II)	0.153	4.18	0.961	0.041	4.59	0.985	
MgAl-LDHs	Cr(VI)	0.040	4.253	0.913	0.021	89.286	1.000	[146]
	Ni(II)	0.032	2.238	0.984	0.013	70.423	1.000	
Two-Dimensional $Ti_3C_2T_x$ MXene	Cu(II)	0.408	1.147	0.24	0.0132	39.22	0.999	[127]

Table 3.
Study of kinetics of heavy metal removal with nanosheets.

Nanosheet	Mechanism	Ref.
Few layered magnetic graphene oxide	The coordination of the lone-pair electrons in C=O- and C–O-related groups toward Cd(II) and Cu(II)	[144]
Boron Nitride	complexation between surface –NH2 functional groups and metal ions, ion exchanges between H of –OH and metal ions, and electrostatic interactions associated with combinations between -O- and metal ions	[148–150]
MoS_2	Reduction-oxidation (redox) reaction	[109]
MgAl-LDHs	The synergistic effect of outer-sphere surface complexes on the adsorbent surface and ion exchange in the interlayer space	[146]
Two-Dimensional $Ti_3C_2T_x$ MXene	The contribution of the surface functional groups	[127]

Table 4.
Study of mechanism of heavy metal removal with nanosheets.

mechanism between nanosheet adsorbents and heavy metals in aqueous solutions but the adsorption of heavy metals mainly involves the electrostatic attraction between the opposite charges, ion exchange, and surface complexation.

4. Conclusion

Nowadays, application of nanosheets for water/wastewater treatment especially for adsorption of heavy metals has attracted more attentions. Because of diversity of nanosheets, which is due to different synthesis methods and novel nanoparticles, studying heavy metal removal improving. Due to their large surface area, porosity, distinct surface charge, surface functionality, and ion adsorption characteristics, nanosheets have strong adsorption capabilities. In this review, mechanism and effective parameters of using nanosheets sorbents in removing heavy metals from aqueous solutions using experimental and simulation methods studied. Synthesis, adsorption mechanism, adsorption performance, and effective parameters of carbon products (graphene), boron nitride materials (BNM), transition metal dichalcogenides (TMDs), layered double hydroxides (LDHs), and MXene as most common 2D nanosheets are studied.

Adsorption is the main mechanism of heavy metal ions removal in 2D nanosheets. Selectivity, adsorption/desorption performance, adsorption capacity, and regeneration experiments are the most characteristics parameters, which researchers have been studying using experiments and simulation methods. The adsorption behavior of nanosheets can improve by functionalizing with magnetic particles. In nanosheets with pores, size and chemistry of the pores exhibit to play an important role in controlling the water flux and ion rejection. For instance, nitrogen on the edges of pores were allowed higher fluxes than boron-lined pores. However, novel synthesis methods could improve the adsorption capability of the serpents. Oppositely charged 2D nanosheets of LDH and –MnO_2 nanosheets assembled in the presence of zwitterionic histidine biomolecules and a novel layer-by-layer LDH/histidine/MnO_2 (LDH/his/MnO_2) nanohybrids were created, which has a massive capacity of Pb(II) and As(V) removal. Novel nanosheets should be designed and prepared with improved adsorption/desorption cycles, easier separation from aqueous solution, non-toxic, and easier production method.

Density functional theory and molecular dynamics, as useful tools for simulation, can develop our insight about interaction between adsorbent and heavy metal ions and effective parameters. The superiority of this method in pre-production simulation has made it possible to rely on the use of this method in the design of new adsorbents.

Isothermal and kinetics models for heavy metal removal were studied, and Langmuir Freundlich models were best fitted to the adsorption mechanism of nanosheets. In addition, pseudo-second order kinetic model showed that chemisorption is the main rate-limiting step of adsorption process. Electrostatic attraction, ion exchange, and surface complexation are the main involving adsorption mechanism through heavy metal removal from aqueous environments.

Combination of experimental approaches and simulations for design of new nanoadsorbents has been growing, and study of interaction between nanosheets and heavy metal ions with consideration of effective parameters including environmental temperature, pH, material types, and heavy metal concentration on thermodynamics resulted of adsorbent mechanism should be organized.

Author details

Mostafa Khosroupour Arabi and Morteza Ghorbanzadeh Ahangari*
Faculty of Engineering and Technology, Department of Mechanical Engineering, University of Mazandaran, Babolsar, Iran

*Address all correspondence to: ghorbanzadeh.morteza@gmail.com

References

[1] Adithya SP, Sidharthan DS, Abhinandan R, Balagangadharan K, Selvamurugan N. Nanosheets-incorporated bio-composites containing natural and synthetic polymers/ceramics for bone tissue engineering. International Journal of Biological Macromolecules. 2020;**164**:1960-1972

[2] Lu H, Wang J, Stoller M, Wang T, Bao Y, Hao H. An overview of nanomaterials for water and wastewater treatment. Advances in Materials Science and Engineering. 2016;**2016**:1-10

[3] Zhang L, Yang L, Chen J, Zhou X. Adsorption of SO_2 and NH_3 onto copper/graphene nanosheets composites: Statistical physics interpretations, thermodynamic investigations, and site energy distribution analyses. Chemical Engineering Journal. 2022;**446**(3):137224

[4] Rahimi R, Solimannejad M, Ehsanfar Z. Potential application of XC3 (X= B, N) nanosheets in drug delivery of hydroxyurea anticancer drug: A comparative DFT study. Molecular Physics. 2022;**120**(6):e2014587

[5] Wang X, Pakdel A, Zhang J, Weng Q, Zhai T, Zhi C, et al. Large-surface-area BN nanosheets and their utilization in polymeric composites with improved thermal and dielectric properties. Nanoscale Research Letters. 2012;**7**(1):1-7

[6] Liu F, Wang X, Chen X, Song X, Tian J, Cui H. Porous ZnO ultrathin nanosheets with high specific surface areas and abundant oxygen vacancies for acetylacetone gas sensing. ACS Applied Materials & Interfaces. 2019;**11**(27):24757-24763

[7] Siddeeg SM, Alsaiari NS, Tahoon MA, Rebah FB. The application of nanomaterials as electrode modifiers for the electrochemical detection of ascorbic acid. International Journal of Electrochemical Science. 2020;**15**:3327-3346

[8] Verma R, Asthana A, Singh AK, Prasad S, Susan MA. Novel glycine-functionalized magnetic nanoparticles entrapped calcium alginate beads for effective removal of lead. Microchemical Journal. 2017;**130**:168-178

[9] Recepoglu YK, Goren AY, Vatanpour V, Yoon Y, Khataee A. Boron carbon nitride nanosheets in water and wastewater treatment: A critical review. Desalination. 2022;**533**:115782

[10] Yaqoob AA, Parveen T, Umar K, Mohamad Ibrahim MN. Role of nanomaterials in the treatment of wastewater: A review. Water. 2020;**12**(2):495

[11] Purkait T, Ahammed R, De Sarkar A, Dey RS. The role of exfoliating solvents for control synthesis of few-layer graphene-like nanosheets in energy storage applications: Theoretical and experimental investigation. Applied Surface Science. 2020 Apr;**15**(509):145375

[12] Mahmood N, De Castro IA, Pramoda K, Khoshmanesh K, Bhargava SK, Kalantar-Zadeh K. Atomically thin two-dimensional metal oxide nanosheets and their heterostructures for energy storage. Energy Storage Materials. 2019;**16**:455-480

[13] Xu L, Wang J. The application of graphene-based materials for the removal of heavy metals and radionuclides from water and wastewater. Critical Reviews in Environmental Science and Technology. 2017;**47**(12):1042-1105

[14] Qasem NA, Mohammed RH, Lawal DU. Removal of heavy metal ions from wastewater: A comprehensive and critical review. Npj Clean Water. 2021;**4**(1):1-5

[15] Ethaib S, Al-Qutaifia S, Al-Ansari N, Zubaidi SL. Function of nanomaterials in removing heavy metals for water and wastewater remediation: A review. Environments. 2022;**9**(10):123

[16] Peng R, Li H, Chen Y, Ren F, Tian F, Gu Y, et al. Highly efficient and selectivity removal of heavy metal ions using single-layer NaxKyMnO2 nanosheet: A combination of experimental and theoretical study. Chemosphere. 2021;**275**:130068

[17] Tahoon MA, Siddeeg SM, Salem Alsaiari N, Mnif W, Ben RF. Effective heavy metals removal from water using nanomaterials: A review. PRO. 2020;**8**(6):645

[18] Mushtaq F, Zahid M, Bhatti IA, Nasir S, Hussain T. Possible applications of coal fly ash in wastewater treatment. Journal of environmental management. 2019;**240**:27-46

[19] Singh NB, Agarwal A, De A, Singh P. Coal fly ash: An emerging material for water remediation. International Journal of Coal Science & Technology. 2022;**9**(1):44

[20] Azam K, Shezad N, Shafiq I, Akhter P, Akhtar F, Jamil F, et al. A review on activated carbon modifications for the treatment of wastewater containing anionic dyes. Chemosphere. 2022;**306**:135566

[21] Rahmani-Sani A, Singh P, Raizada P, Lima EC, Anastopoulos I, Giannakoudakis DA, et al. Use of chicken feather and eggshell to synthesize a novel magnetized activated carbon for sorption of heavy metal ions. Bioresource Technology. 2020;**297**:122452

[22] Belova TP. Adsorption of heavy metal ions (Cu2+, Ni2+, Co2+ and Fe2+) from aqueous solutions by natural zeolite. Heliyon. 2019;**5**(9):e02320

[23] Ugwu EI, Othmani A, Nnaji CC. A review on zeolites as cost-effective adsorbents for removal of heavy metals from aqueous environment. International Journal of Environmental Science and Technology. 2022;**19**(8):8061-8084

[24] Tasić ŽZ, Bogdanović GD, Antonijević MM. Application of natural zeolite in wastewater treatment: A review. Journal of Mining and Metallurgy A: Mining. 2019;**55**(1):67-79

[25] Hızal J, Yılmazoğlu M. Montmorillonite Clay Composite for Heavy Metal Removal from Water. Green Adsorbents to Remove Metals: Dyes and Boron from Polluted Water; 2021. pp. 93-112

[26] Bhat AH, Rangreez TA, Chisti HT. Wastewater treatment and biomedical applications of montmorillonite based nanocomposites: A review. Current Analytical Chemistry. 2022;**18**(3):269-287

[27] Liu L, Zhang C, Jiang W, Li X, Dai Y, Jia H. Understanding the sorption behaviors of heavy metal ions in the interlayer and nanopore of montmorillonite: A molecular dynamics study. Journal of Hazardous Materials. 2021;**416**:125976

[28] Begum S, Yuhana NY, Saleh NM, Kamarudin NH, Sulong AB. Review of chitosan composite as a heavy metal adsorbent: Material preparation and properties. Carbohydrate Polymers. 2021;**259**:117613

[29] Upadhyay U, Sreedhar I, Singh SA, Patel CM, Anitha KL. Recent advances in

heavy metal removal by chitosan based adsorbents. Carbohydrate Polymers. 2021;**251**:117000

[30] Sarode S, Upadhyay P, Khosa MA, Mak T, Shakir A, Song S, et al. Overview of wastewater treatment methods with special focus on biopolymer chitin-chitosan. International Journal of Biological Macromolecules. 2019;**121**:1086-1100

[31] Yang J, Hou B, Wang J, Tian B, Bi J, Wang N, et al. Nanomaterials for the removal of heavy metals from wastewater. Nanomaterials. 2019;**9**(3):424

[32] Sharma YC, Srivastava V, Singh VK, Kaul SN, Weng CH. Nano-adsorbents for the removal of metallic pollutants from water and wastewater. Environmental Technology. 2009;**30**(6):583-609

[33] Fu L, Yan Z, Zhao Q, Yang H. Novel 2D nanosheets with potential applications in heavy metal purification: A review. Advanced Materials Interfaces. 2018;**5**(23):1801094

[34] Loske L, Nakagawa K, Yoshioka T, Matsuyama H. 2D nanocomposite membranes: Water purification and fouling mitigation. Membranes. 2020;**10**(10):295

[35] Fatima J, Shah AN, Tahir MB, Mehmood T, Shah AA, Tanveer M, et al. Tunable 2D nanomaterials; their key roles and mechanisms in water purification and monitoring. Frontiers in Environmental Science. 2022;**10**:210

[36] Mashhadzadeh AH, Fathalian M, Ahangari MG, Shahavi MH. DFT study of Ni, Cu, Cd and Ag heavy metal atom adsorption onto the surface of the zinc-oxide nanotube and zinc-oxide graphene-like structure. Materials Chemistry and Physics. 2018;**220**:366-373

[37] Rassoulinejad-Mousavi SM, Azamat J, Khataee A, Zhang Y. Molecular dynamics simulation of water purification using zeolite MFI nanosheets. Separation and Purification Technology. 2020;**234**:116080

[38] Zhan G, Li J, Hu Y, Zhao S, Cao S, Jia F, et al. The surface hydroxyl and oxygen vacancy dependent Cr (VI) adsorption performance of BiOCl. Environmental Science: Nano. 2020;**7**(5):1454-1463

[39] Zhang J, Liao J, Yang F, Xu M, Lin S. Regulation of the electroanalytical performance of ultrathin titanium dioxide nanosheets toward lead ions by non-metal doping. Nanomaterials. 2017;**7**(10):327

[40] Díaz-Rodríguez TG, Pacio M, Agustín-SerranoR,Juárez-SantiestebanH, Muñiz J. Understanding structure of small TiO_2 nanoparticles and adsorption mechanisms of PbS quantum dots for solid-state applications: A combined theoretical and experimental study. Theoretical Chemistry Accounts. 2019;**138**(7):92

[41] Luan J, Zhao C, Zhai Q, Liu W, Ke X, Liu X. The highly efficient simultaneous removal of Pb2+ and methylene blue induced by the release of endogenous active sites of montmorillonite. Water Science & Technology. 2022;**86**(9):2336-2347

[42] Niu J, Zhang W, Li S, Yan W, Hao X, Wang Z, et al. An electroactive montmorillonite/polyaniline nanocomposite film: Superfast ion transport and ultra-affinity ion recognition for rapid and selective separation of Pb2+ ions. Chemical Engineering Journal. 2021;**413**:127750

[43] Zhou WY, Liu JY, Song JY, Li JJ, Liu JH, Huang XJ. Surface-electronic-state-modulated, single-crystalline

(001) TiO_2 nanosheets for sensitive electrochemical sensing of heavy-metal ions. Analytical Chemistry. 2017;**89**(6):3386-3394

[44] Zhou WY, Li SS, Song JY, Jiang M, Jiang TJ, Liu JY, et al. High electrochemical sensitivity of TiO_2–x nanosheets and an electron-induced mutual interference effect toward heavy metal ions demonstrated using X-ray absorption fine structure spectra. Analytical Chemistry. 2018;**90**(7):4328-4337

[45] Li H, Zhang L. Oxygen vacancy induced selective silver deposition on the {001} facets of BiOCl single-crystalline nanosheets for enhanced Cr (VI) and sodium pentachlorophenate removal under visible light. Nanoscale. 2014;**6**(14):7805-7810

[46] Xu X, Zeng J, Wu Y, Wang Q, Wu S, Gu H. Preparation and application of graphene–based materials for heavy metal removal in tobacco industry: A review. Separations. 2022;**9**(12):401

[47] Park YG, Nam SN, Jang M, Park CM, Her N, Sohn J, et al. Boron nitride-based nanomaterials as adsorbents in water: A review. Separation and Purification Technology. 2022;**288**:120637

[48] Su Y, Liu D, Yang G, Han Q, Qian Y, Liu Y, et al. Transition metal dichalcogenide (TMD) membranes with ultrasmall nanosheets for ultrafast molecule separation. ACS Applied Materials & Interfaces. 2020;**12**(40):45453-45459

[49] Huang H, Zha J, Li S, Tan C. Two-dimensional alloyed transition metal dichalcogenide nanosheets: Synthesis and applications. Chinese Chemical Letters. 2022;**33**(1):163-176

[50] Liu W, Liu Y, Yuan Z, Lu C. Recent advances in the detection and removal of heavy metal ions using functionalized layered double hydroxides: A review. Industrial Chemistry & Materials. 2023;**1**:79-92

[51] Guan X, Yuan X, Zhao Y, Wang H, Wang H, Bai J, et al. Application of functionalized layered double hydroxides for heavy metal removal: A review. Science of the Total Environment. 2022;**838**:155693

[52] Dong Y, Kong X, Luo X, Wang H. Adsorptive removal of heavy metal anions from water by layered double hydroxide: A review. Chemosphere. 2022;**303**:134685

[53] Othman Z, Mackey HR, Mahmoud KA. A critical overview of MXenes adsorption behavior toward heavy metals. Chemosphere. 2022;**295**:133849

[54] Zhao G, Li J, Ren X, Chen C, Wang X. Few-layered graphene oxide nanosheets as superior sorbents for heavy metal ion pollution management. Environmental Science & Technology. 2011;**45**(24):10454-10462

[55] Yang K, Chen B, Zhu X, Xing B. Aggregation, adsorption, and morphological transformation of graphene oxide in aqueous solutions containing different metal cations. Environmental Science & Technology. 2016;**50**(20):11066-11075

[56] Zhang L, Hu X, Zhou Q. Sunlight-assisted tailoring of surface nanostructures on single-layer graphene nanosheets for highly efficient cation capture and high-flux desalination. Carbon. 2020;**161**:674-684

[57] Mirzaei M, Mohammadi T, Kasiri N, Tofighy MA. Fabrication of magnetic field induced mixed matrix membranes containing GO/Fe_3O_4 nanohybrids with enhanced antifouling properties

for wastewater treatment applications. Journal of Environmental Chemical Engineering. 2021;**9**(4):105675

[58] Katubi KM, Alsaiari NS, Alzahrani FMM, Siddeeg SA, Tahoon M. Synthesis of manganese ferrite/graphene oxide magnetic nanocomposite for pollutants removal from water. PRO. 2021;**9**(4):589

[59] Dai K, Liu G, Xu W, Deng Z, Wu Y, Zhao C, et al. Judicious fabrication of bifunctionalized graphene oxide/ $MnFe_2O_4$ magnetic nanohybrids for enhanced removal of Pb (II) from water. Journal of Colloid and Interface Science. 2020;**579**:815-822

[60] Hu XJ, Liu YG, Wang H, Chen AW, Zeng GM, Liu SM, et al. Removal of Cu (II) ions from aqueous solution using sulfonated magnetic graphene oxide composite. Separation and Purification Technology. 2013;**108**:189-195

[61] Wei MP, Chai H, Cao YL, Jia DZ. Sulfonated graphene oxide as an adsorbent for removal of Pb2+ and methylene blue. Journal of Colloid and Interface Science. 2018;**524**:297-305

[62] Zhou G, Xu X, Zhu W, Feng B, Hu J. Dispersedly embedded loading of Fe_3O_4 nanoparticles into graphene nanosheets for highly efficient and recyclable removal of heavy metal ions. New Journal of Chemistry. 2015;**39**(9):7355-7362

[63] Li J, Zhang S, Chen C, Zhao G, Yang X, Li J, et al. Removal of Cu (II) and fulvic acid by graphene oxide nanosheets decorated with Fe_3O_4 nanoparticles. ACS Applied Materials & Interfaces. 2012;**4**(9):4991-5000

[64] Peer FE, Bahramifar N, Younesi H. Removal of Cd (II), Pb (II) and Cu (II) ions from aqueous solution by polyamidoamine dendrimer grafted magnetic graphene oxide nanosheets. Journal of the Taiwan Institute of Chemical Engineers. 2018;**87**:225-240

[65] Xu H, Zhang S, Zhang T, Huang W, Dai Y, Zheng R, et al. An electrochemiluminescence biosensor for cadmium ion based on target-induced strand displacement amplification and magnetic Fe_3O_4-GO nanosheets. Talanta. 2022;**237**:122967

[66] Wu Y, Li Z, Chen J, Yu C, Huang X, Zhao C, et al. Graphene nanosheets decorated with tunable magnetic nanoparticles and their efficiency of wastewater treatment. Materials Research Bulletin. 2015;**68**:234-239

[67] Najafabadi HH, Irani M, Rad LR, Haratameh AH, Haririan I. Removal of Cu 2+, Pb 2+ and Cr 6+ from aqueous solutions using a chitosan/graphene oxide composite nanofibrous adsorbent. Rsc Advances. 2015;**5**(21):16532-16539

[68] Menazea AA, Ezzat HA, Omara W, Basyouni OH, Ibrahim SA, Mohamed AA, et al. Chitosan/graphene oxide composite as an effective removal of Ni, Cu, As, Cd and Pb from wastewater. Computational and Theoretical Chemistry. 2020;**1189**:112980

[69] Zhao C, Luan J, Zhai Q, Liu W, Ge H, Ke X, et al. Releasing SiO tetrahedron and AlO octahedron from montmorillonite to enhance the adsorption performance of carbon@ chitosan@ montmorillonite nanosheet for cationic dyes: Coupling quantum chemistry simulations with experiments. Science of the Total Environment. 2022;**851**:158174

[70] Jiang T, Liu W, Mao Y, Zhang L, Cheng J, Gong M, et al. Adsorption behavior of copper ions from aqueous solution onto graphene oxide–CdS

composite. Chemical Engineering Journal. 2015;**259**:603-610

[71] Huang R, Ma X, Li X, Guo L, Xie X, Zhang M, et al. A novel ion-imprinted polymer based on graphene oxide-mesoporous silica nanosheet for fast and efficient removal of chromium (VI) from aqueous solution. Journal of Colloid and Interface Science. 2018;**514**:544-553

[72] Zhang L, Song F, Wang S, Wang H, Yang W, Li Y. Efficient removal of hexavalent chromium and Congo red by graphene oxide/silica nanosheets with multistage pores. Journal of Chemical & Engineering Data. 2020;**65**(9):4354-4368

[73] Liu M, Xu J, Cheng B, Ho W, Yu J. Synthesis and adsorption performance of Mg (OH) 2 hexagonal nanosheet–graphene oxide composites. Applied Surface Science. 2015;**332**:121-129

[74] Bukhari A, Ijaz I, Zain H, Mehmood U, Iqbal MM, Gilani E, et al. Introduction of CdO nanoparticles into graphene and graphene oxide nanosheets for increasing adsorption capacity of Cr from wastewater collected from petroleum refinery. Arabian Journal of Chemistry. 2023;**16**(2):104445

[75] Liu R, Zhou J, Gao X, Li J, Xie Z, Li Z, et al. Graphdiyne filter for decontaminating lead-ion-polluted water. Advanced Electronic Materials. 2017;**3**(11):1700122

[76] Baghbani NB, Azamat J, Erfan-Niya H, Majidi S, Khazini L. Molecular insights into water desalination performance of pristine graphdiyne nanosheet membrane. Journal of Molecular Graphics and Modelling. 2020;**101**:107729

[77] Yu H, Xue Y, Li Y. Graphdiyne and its assembly architectures: Synthesis, functionalization, and applications. Advanced Materials. 2019;**31**(42):1803101

[78] Mashhadzadeh AH, Vahedi AM, Ardjmand M, Ahangari MG. Investigation of heavy metal atoms adsorption onto graphene and graphdiyne surface: A density functional theory study. Superlattices and Microstructures. 2016;**100**:1094-1102

[79] Alija A, Gashi D, Plakaj R, Omaj A, Thaçi V, Reka A, et al. A theoretical and experimental study of the adsorptive removal of hexavalent chromium ions using graphene oxide as an adsorbent. Open Chemistry. 2020;**18**(1):936-942

[80] Kommu A, Namsani S, Singh JK. Removal of heavy metal ions using functionalized graphene membranes: A molecular dynamics study. RSC Advances. 2016;**6**(68):63190-63199

[81] Li Y, Xu Z, Liu S, Zhang J, Yang X. Molecular simulation of reverse osmosis for heavy metal ions using functionalized nanoporous graphenes. Computational Materials Science. 2017;**139**:65-74

[82] Azamat J, Sattary BS, Khataee A, Joo SW. Removal of a hazardous heavy metal from aqueous solution using functionalized graphene and boron nitride nanosheets: Insights from simulations. Journal of Molecular Graphics and Modelling. 2015;**61**:13-20

[83] Azamat J. Removal of nickel (II) from aqueous solution by graphene and boron nitride nanosheets. Journal of Water and Environmental Nanotechnology. 2017;**2**(1):26-33

[84] Lohrasebi A, Rikhtehgaran S. Ion separation and water purification by applying external electric field on porous graphene membrane. Nano Research. 2018;**11**(4):2229-2236

[85] Khnifira M, Boumya W, Attarki J, Mahsoune A, Sadiq M, Abdennouri M, et al. A combined DFT, Monte Carlo,

and MD simulations of adsorption study of heavy metals on the carbon graphite (111) surface. Chemical Physics Impact. 2022;**5**:100121

[86] Majidi S, Erfan-Niya H, Azamat J, Cruz-Chú ER, Walther JH. The performance of a C2N membrane for heavy metal ions removal from water under external electric field. Separation and Purification Technology. 2022;**289**:120770

[87] Wang H, Wang W, Wang H, Zhang F, Li Y, Fu Z. Urchin-like boron nitride hierarchical structure assembled by nanotubes-nanosheets for effective removal of heavy metal ions. Ceramics International. 2018;**44**(11):12216-12224

[88] Wang S, Jia F, Kumar P, Zhou A, Hu L, Shao X, et al. Hierarchical porous boron nitride nanosheets with versatile adsorption for water treatment. Colloids and Surfaces A: Physicochemical and Engineering Aspects. 2020;**598**:124865

[89] Liu F, Yu J, Ji X, Qian M. Nanosheet-structured boron nitride spheres with a versatile adsorption capacity for water cleaning. ACS Applied Materials & Interfaces. 2015;**7**(3):1824-1832

[90] Liu T, Li Y, He J, Zhang K, Hu Y, Chen X, et al. Few-layered boron nitride nanosheets as superior adsorbents for the rapid removal of lead ions from water. Journal of Materials Science. 2019;**54**(7):5366-5380

[91] Li J, Jin P, Dai W, Wang C, Li R, Wu T, et al. Excellent performance for water purification achieved by activated porous boron nitride nanosheets. Materials Chemistry and Physics. 2017;**196**:186-193

[92] Peng D, Jiang W, Li FF, Zhang L, Liang RP, Qiu JD. One-pot synthesis of boron carbon nitride nanosheets for facile and efficient heavy metal ions removal. ACS Sustainable Chemistry & Engineering. 2018;**6**(9):11685-11694

[93] Krishna Kumar AS, Warchol J, Matusik J, Tseng WL, Rajesh N, Bajda T. Heavy metal and organic dye removal via a hybrid porous hexagonal boron nitride-based magnetic aerogel. NPJ Clean Water. 2022;**5**(1):24

[94] Bangari RS, Yadav VK, Singh JK, Sinha N. Fe_3O_4-functionalized boron nitride nanosheets as novel adsorbents for removal of arsenic (III) from contaminated water. ACS Omega. 2020;**5**(18):10301-10314

[95] Bangari RS, Singh AK, Namsani S, Singh JK, Sinha N. Magnetite-coated boron nitride nanosheets for the removal of arsenic (V) from water. ACS Applied Materials & Interfaces. 2019;**11**(21):19017-19028

[96] Gao H, Shi Q, Rao D, Zhang Y, Su J, Liu Y, et al. Rational design and strain engineering of nanoporous boron nitride nanosheet membranes for water desalination. The Journal of Physical Chemistry C. 2017;**121**(40):22105-22113

[97] Azamat J, Khataee A, Joo SW. Separation of copper and mercury as heavy metals from aqueous solution using functionalized boron nitride nanosheets: A theoretical study. Journal of Molecular Structure. 2016;**1108**:144-149

[98] Azamat J, Sardroodi JJ, Poursoltani L, Jahanshahi D. Functionalized boron nitride nanosheet as a membrane for removal of Pb2+ and Cd2+ ions from aqueous solution. Journal of Molecular Liquids. 2021;**321**:114920

[99] Srivastava R, Kommu A, Sinha N, Singh JK. Removal of arsenic ions using hexagonal boron nitride and graphene

nanosheets: A molecular dynamics study. Molecular Simulation. 2017;**43**(13-16):985-996

[100] Yadav A, Dindorkar SS, Ramisetti SB, Sinha N. Simultaneous adsorption of methylene blue and arsenic on graphene, boron nitride and boron carbon nitride nanosheets: Insights from molecular simulations. Journal of Water Process Engineering. 2022;**46**:102653

[101] Saias E, Ismach A, Zucker I. Engineering the performance and stability of molybdenum Disulfide for heavy metal removal. ACS Applied Materials & Interfaces. 2022;**15**(5):6603-6611

[102] Wang Z, Sim A, Urban JJ, Mi B. Removal and recovery of heavy metal ions by two-dimensional MoS_2 nanosheets: Performance and mechanisms. Environmental Science & Technology. 2018;**52**(17):9741-9748

[103] Huang S, You Z, Jiang Y, Zhang F, Liu K, Liu Y, et al. Fabrication of ultrathin MoS2 nanosheets and application on adsorption of organic pollutants and heavy metals. PRO. 2020;**8**(5):504

[104] Sun H, Wu T, Zhang Y, Ng DH, Wang G. Structure-enhanced removal of Cr (vi) in aqueous solutions using MoS_2 ultrathin nanosheets. New Journal of Chemistry. 2018;**42**(11):9006-9015

[105] Ai K, Ruan C, Shen M, Lu L. MoS_2 nanosheets with widened interlayer spacing for high-efficiency removal of mercury in aquatic systems. Advanced Functional Materials. 2016;**26**(30):5542-5549

[106] Wang J, Zhang W, Yue X, Yang Q, Liu F, Wang Y, et al. One-pot synthesis of multifunctional magnetic ferrite–MoS_2–carbon dot nanohybrid adsorbent for efficient Pb (ii) removal. Journal of Materials Chemistry A. 2016;**4**(10):3893-3900

[107] Nie G, Qiu S, Wang X, Du Y, Zhang Q, Zhang Y, et al. A millimeter-sized negatively charged polymer embedded with molybdenum disulfide nanosheets for efficient removal of Pb (II) from aqueous solution. Chinese Chemical Letters. 2021;**32**(7):2342-2346

[108] Wang Z, Tu Q, Sim A, Yu J, Duan Y, Poon S, et al. Superselective removal of lead from water by two-dimensional MoS_2 nanosheets and layer-stacked membranes. Environmental Science & Technology. 2020;**54**(19):12602-12611

[109] Luo J, Fu K, Sun M, Yin K, Wang D, Liu X, et al. Phase-mediated heavy metal adsorption from aqueous solutions using two-dimensional layered MoS_2. ACS Applied Materials & Interfaces. 2019;**11**(42):38789-38797

[110] Mu XL, Gao X, Zhao HT, George M, Wu T. Density functional theory study of the adsorption of elemental mercury on a 1T-MoS2 monolayer. Journal of Zhejiang University-Science A. 2018;**19**(1):60-67

[111] Yi H, Zhang X, Jia F, Wei Z, Zhao Y, Song S. Competition of Hg2+ adsorption and surface oxidation on MoS2 surface as affected by sulfur vacancy defects. Applied Surface Science. 2019;**483**:521-528

[112] Azamat J, Khataee A. Improving the performance of heavy metal separation from water using MoS2 membrane: Molecular dynamics simulation. Computational Materials Science. 2017;**137**:201-207

[113] Yuan X, Wang Y, Wang J, Zhou C, Tang Q, Rao X. Calcined graphene/MgAl-layered double hydroxides for enhanced Cr (VI) removal. Chemical Engineering Journal. 2013;**221**:204-213

[114] Gherca D, Borhan AI, Mihai MM, Herea DD, Stoian G, Roman T, et al. Magnetite-induced topological transformation of 3D hierarchical MgAl layered double hydroxides to highly dispersed 2D magnetic hetero-nanosheets for effective removal of cadmium ions from aqueous solutions. Materials Chemistry and Physics. 2022;**284**:126047

[115] Shamsayei M, Yamini Y, Asiabi H. Layer-by-layer assembly of layered double hydroxide/histidine/δ-MnO_2 nanosheets: Synthesis, characterization, and applications. Applied Clay Science. 2020;**188**:105540

[116] Li J, Huang Q, Yu H, Yan L. Enhanced removal performance and mechanistic study of Cu2+, Cd2+, and Pb2+ by magnetic layered double hydroxide nanosheets assembled on graphene oxide. Journal of Water Process Engineering. 2022;**48**:102893

[117] Wang H, Wang S, Chen Z, Zhou X, Wang J, Chen Z. Engineered biochar with anisotropic layered double hydroxide nanosheets to simultaneously and efficiently capture Pb2+ and CrO42− from electroplating wastewater. Bioresource Technology. 2020;**306**:123118

[118] Wang J, Kong X, Yang M, Xiong W, Li Z, Zhou H, et al. Superstable mineralization of heavy metals using low-cost layered double hydroxide nanosheets: Toward water remediation and soil fertility enhancement. Industrial & Engineering Chemistry Research. 2022;**1**(2):20210052

[119] Zahir MH, Irshad K, Rahman MM, Shaikh MN, Rahman MM. Efficient capture of heavy metal ions and arsenic with a CaY–carbonate layered double-hydroxide nanosheet. ACS Omega. 2021;**6**(35):22909-22921

[120] Ali J, Wang H, Ifthikar J, Khan A, Wang T, Zhan K, et al. Efficient, stable and selective adsorption of heavy metals by thio-functionalized layered double hydroxide in diverse types of water. Chemical Engineering Journal. 2018;**332**:387-397

[121] Wang Y, Gu Y, Xie D, Qin W, Zhang H, Wang G, et al. A hierarchical hybrid monolith: MoS 4 2−−intercalated NiFe layered double hydroxide nanosheet arrays assembled on carbon foam for highly efficient heavy metal removal. Journal of Materials Chemistry A. 2019;**7**(20):12869-12881

[122] Zhu S, Chen Y, Khan MA, Xu H, Wang F, Xia M. In-depth study of heavy metal removal by an etidronic acid-functionalized layered double hydroxide. ACS Applied Materials & Interfaces. 2022;**14**(5):7450-7463

[123] Zhu S, Khan MA, Wang F, Bano Z, Xia M. Rapid removal of toxic metals Cu2+ and Pb2+ by amino trimethylene phosphonic acid intercalated layered double hydroxide: A combined experimental and DFT study. Chemical Engineering Journal. 2020;**392**:123711

[124] Shahzad A, Rasool K, Miran W, Nawaz M, Jang J, Mahmoud KA, et al. Mercuric ion capturing by recoverable titanium carbide magnetic nanocomposite. Journal of Hazardous Materials. 2018;**344**:811-818

[125] Shahzad A, Nawaz M, Moztahida M, Jang J, Tahir K, Kim J, et al. Ti3C2Tx MXene core-shell spheres for ultrahigh removal of mercuric ions. Chemical Engineering Journal. 2019;**368**:400-408

[126] Fu K, Liu X, Yu D, Luo J, Wang Z, Crittenden JC. Highly efficient and selective Hg (II) removal from water using multilayered Ti3C2O x MXene via adsorption coupled with

catalytic reduction mechanism. Environmental Science & Technology. 2020;**54**(24):16212-16220

[127] Asif S, Kashif R, Waheed M, Mohsin N, Jiseon J, Sung LD. Two-dimensional Ti3C2Tx MXene nanosheets for efficient copper removal from water. ACS Sustainable Chemistry & Engineering. 2017;**5**(12):11481-11488

[128] Guo X, Zhang X, Zhao S, Huang Q, Xue J. High adsorption capacity of heavy metals on two-dimensional MXenes: An ab initio study with molecular dynamics simulation. Physical Chemistry Chemical Physics. 2016;**18**(1):228-233

[129] Chadha U, Selvaraj SK, Thanu SV, Cholapadath V, Abraham AM, Manoharan M, et al. A review of the function of using carbon nanomaterials in membrane filtration for contaminant removal from wastewater. Materials Research Express. 2022;**9**(1):012003

[130] Antipina LY, Kotyakova KY, Tregubenko MV, Shtansky DV. Experimental and theoretical study of sorption capacity of hexagonal boron nitride nanoparticles: Implication for wastewater purification from antibiotics. Nanomaterials. 2022;**12**(18):3157

[131] Avasarala S, Bose S. 2D nanochannels and huge specific surface area offer unique ways for water remediation and adsorption: Assessing the strengths of hexagonal boron nitride in separation technology. Functional Composite Materials. 2023;**4**(1):5

[132] Smith KK, Redeker ND, Rios JC, Mecklenburg MH, Marcischak JC, Guenthner AJ, et al. Surface modification and functionalization of boron nitride nanotubes via condensation with saturated and unsaturated alcohols for high performance polymer composites. ACS Applied Nano Materials. 2019;**2**(7):4053-4060

[133] Shin H, Guan J, Zgierski MZ, Kim KS, Kingston CT, Simard B. Covalent functionalization of boron nitride nanotubes via reduction chemistry. ACS Nano. 2015;**9**(12):12573-12582

[134] Zhao F, Peydayesh M, Ying Y, Mezzenga R, Ping J. Transition metal dichalcogenide–silk nanofibril membrane for one-step water purification and precious metal recovery. ACS Applied Materials & Interfaces. 2020;**12**(21):24521-24530

[135] Peng H, Wang R, Mei L, Zhang Q, Ying T, Qian Z, et al. Transition metal dichalcogenide-based functional membrane: Synthesis, modification, and water purification applications. Matter. 2023;**6**(1):59-96

[136] Asif MB, Iftekhar S, Maqbool T, Pramanik BK, Tabraiz S, Sillanpää M, et al. Two-dimensional nanoporous and lamellar membranes for water purification: Reality or a myth? Chemical Engineering Journal. 2022;**432**:134335

[137] Tang Z, Qiu Z, Lu S, Shi X. Functionalized layered double hydroxide applied to heavy metal ions absorption: A review. Nanotechnology Reviews. 2020;**9**(1):800-819

[138] Soltani R, Pelalak R, Pishnamazi M, Marjani A, Sarkar SM, Albadarin AB, et al. Novel bimodal micro-mesoporous Ni50Co50-LDH/UiO-66-NH2 nanocomposite for Tl (I) adsorption. Arabian Journal of Chemistry. 2021;**14**(4):103058

[139] Ibrahim Y, Kassab A, Eid K, Abdullah AM, Ozoemena KI, Elzatahry A. Unveiling fabrication and environmental remediation of MXene-based nanoarchitectures in toxic metals removal from wastewater:

Strategy and mechanism. Nanomaterials. 2020;**10**(5):885

[140] Hussain A, Madan S, Madan R. Removal of Heavy Metals from Wastewater by Adsorption [Internet]. Heavy Metals - Their Environmental Impacts and Mitigation. IntechOpen; 2021. Available from: http://dx.doi.org/10.5772/intechopen.95841

[141] Gupta A, Sharma V, Sharma K, Kumar V, Choudhary S, Mankotia P, et al. A review of adsorbents for heavy metal decontamination: Growing approach to wastewater treatment. Materials. 2021;**14**(16):4702

[142] Aigbe UO, Osibote OA. Carbon derived nanomaterials for the sorption of heavy metals from aqueous solution: A review. Environmental Nanotechnology, Monitoring & Management. 2021;**16**:100578

[143] Ahmad SZ, Salleh WN, Ismail AF, Yusof N, Yusop MZ, Aziz F. Adsorptive removal of heavy metal ions using graphene-based nanomaterials: Toxicity, roles of functional groups and mechanisms. Chemosphere. 2020;**248**:126008

[144] Guo T, Bulin C, Ma Z, Li B, Zhang Y, Zhang B, et al. Mechanism of Cd (II) and Cu (II) adsorption onto few-layered magnetic graphene oxide as an efficient adsorbent. ACS Omega. 2021;**6**(25):16535-16545

[145] LI L, GUO X, JIN Y, CHEN C, Abdullah M A, Hadi M M, et al. Distinguished Cd(II) Capture with Rapid and Superior Ability using Porous Hexagonal Boron Nitride: Kinetic and Thermodynamic Aspects[J]. Journal of Inorganic Materials. 2020;**35**(3):284-292

[146] Li J, Cui H, Song X, Zhang G, Wang X, Song Q, et al. Adsorption and intercalation of organic pollutants and heavy metal ions into MgAl-LDHs nanosheets with high capacity. Rsc Advances. 2016;**6**(95):92402-92410

[147] Akpomie KG, Conradie J, Adegoke KA, Oyedotun KO, Ighalo JO, Amaku JF, et al. Adsorption mechanism and modeling of radionuclides and heavy metals onto ZnO nanoparticles: A review. Applied Water Science. 2023;**13**(1):20

[148] Zhu Q, Li Z. Hydrogel-supported nanosized hydrous manganese dioxide: Synthesis, characterization, and adsorption behavior study for Pb2+, Cu2+, Cd2+ and Ni2+ removal from water. Chemical Engineering Journal. 2015;**281**:69-80

[149] Xie Y, Yuan X, Wu Z, Zeng G, Jiang L, Peng X, et al. Adsorption behavior and mechanism of Mg/Fe layered double hydroxide with Fe_3O_4-carbon spheres on the removal of Pb (II) and Cu (II). Journal of Colloid and Interface Science. 2019;**536**:440-455

[150] Zhao X, Wang Y, Wu H, Fang L, Liang J, Fan Q, et al. Insights into the effect of humic acid on Ni (II) sorption mechanism on illite: Batch, XPS and EXAFS investigations. Journal of Molecular Liquids. 2017;**248**:1030-1038

Chapter 5

Preparation of Siloxene-Graphene 2D/2D Heterostructures for High-Performance Supercapacitors in Electric Vehicles

Karthikeyan Krishnamoorthy, Parthiban Pazhamalai, Rajavarman Swaminathan and Sang-Jae Kim

Abstract

The development of wide temperature tolerance supercapacitors (SCs) with high specific energy without compromising specific power is an area of emerging interest owing to the increasing demands for electrochemical energy storage system (EES). This chapter discusses the preparation of siloxene-graphene (rGO) 2D/2D heterostructures (via chemical methods) and examines their potential utility toward SCs for electric vehicles (EVs). The electrochemical characterization of the siloxene-rGO SC showed that they possess high specific energy (55.79 Wh kg^{-1}), and specific power (15, 000 W kg^{-1}). And their ability to operate over a wide temperature range (−15 to 80°C), ensuring their suitability as an EES in EVs. The additional experimental studies suggested the ability of the solar-charged siloxene-rGO SC to drive an electric car, and it can capture the regenerative braking energy during the braking process. This chapter provides a new avenue toward the use of siloxene-rGO SC as a suitable EES for next-generation EVs.

Keywords: siloxene-graphene, 2D/2D heterostructures, supercapacitors for electric vehicles, regenerative braking, solar-charging supercapacitor

1. Introduction

The demands for the development of electric vehicles (EVs) are very rapidly increasing due to the depletion of fossil fuels worldwide [1]. Notably, the use of EVs instead of conventional system possesses the merits of zero carbon emission, thereby reduces environmental pollution [2]. Therefore, electric energy harvesting from renewable methods such as solar, wind, and hydro-power as well as developing high-performance electrochemical energy storage system (EES) are required to develop highly efficient EVs [3–5]. To date, Li-ion batteries (LIBs) are considered as a suitable choice for EES for EVs and other electronic devices owing to their high energy density [6, 7]. However, the major drawback/issues of LIBs for EVs application are their thermal degradation issues (that results in loss of specific capacity) and the deterioration of battery electrode due to the flow of reverse current from the electric motor during regenerative braking or deceleration of EVs [8, 9]. Additionally, the power density of LIBs is low, which limits their ability for rapid acceleration. In this scenario, the supercapacitors (SCs) are proposed as an alternate EES system wherever peak power demands are required such as storage of braking energy regenerated in electric vehicles (EVs), elevators, etc., [10, 11]. The merits or specific features of SCs such as high specific power, and long cycle-life made them as an attractive candidate for not only EVs (new EVs and hybrid EVs) but even as a maintenance-free EES for the wind-patch control systems [12].

Recent research has been focused on developing high-performance SCs to gain energy density to reach the level of battery without reducing their power density [10, 13]. Herein, the discovery of graphene sheets resulted in the achievement of high capacitance surpassing the previous generation carbon such as activated carbon, carbon nanotubes, fullerenes that mainly originated from the two-dimensional (2D) structures [14], high surface-to-volume ratio [15], and their intrinsic high electrical conductivity [16]. The emergence of graphene and its attractive energy storage properties, researchers focused on studying the energy storage properties of various 2D materials from the transition metal dichalcogenide (MoS_2, $MoSe_2$, and WS_2) [17–19] MXenes (2D metal carbides, metal nitrides, and metal carbonitrides) [20–22], metalenes (boron, antimonene, and phosphorene) [23–25], and from silicon family such as siloxene, and their derivatives [26, 27], respectively. Benefitting from the structural merits (for ion-intercalation/de-intercalation) [28], presence of transition metal (contributing electrochemical redox reactions) [29], in-plane electrical conductivity [30], these materials showed better electrochemical charge-storage performance than their bulk counterparts and other morphologies [31]. Herein, it is worth mentioning that siloxene sheets and their derivatives possess maximal power density than the reported silicon-based SCs signifying their applications in on-chip SCs for convenient integration with silicon technology [32]. However, the mechanism of charge-storage in these materials arises from ion-intercalation-mediated capacitance that significantly differs from the electric double-layer capacitance (EDLC) seen in graphene sheets [33, 34]. The major limitation of utilizing these 2D materials as electrodes of SCs in commercialization aspects is their ability to re-stack with the adjacent sheets, thereby results in a capacitance fading effect with an increase in cycle life [35]. To solve this issue, various strategies, like growing nanocrystals of 2D sheets, and developing binder-free electrode configurations directly grown on current collectors, have been investigated recently [14, 36].

In this chapter, we are discussing about the synthesis of siloxene-graphene(rGO) (2D/2D) heterostructures as a novel electrode for SCs that is capable of harvesting regenerative braking energy from EVs. Hybridizing siloxene sheets with graphene sheets results not only in the prevention of re-stacking of individual sheets but also offers two dissimilar charge-storage properties in the fabricated electrode for gaining much better capacitance and energy density without losing their specific power.

2. Results and discussion

Figure 1 schemes the synthesis of siloxene-rGO 2D/2D nanohybrids using a multi-step chemical reaction. The calcium silicide ($CaSi_2$) powders were immersed in an ice-cold HCl solution in which graphene oxide (GO) powders were already predispersed. The reaction is carried out for nearly 4 days which result in the production of siloxene layers via the de-intercalation of calcium ions by releasing hydrogen bubbles [32], which is responsible for the partial reduction of GO into rGO sheets. Finally, the

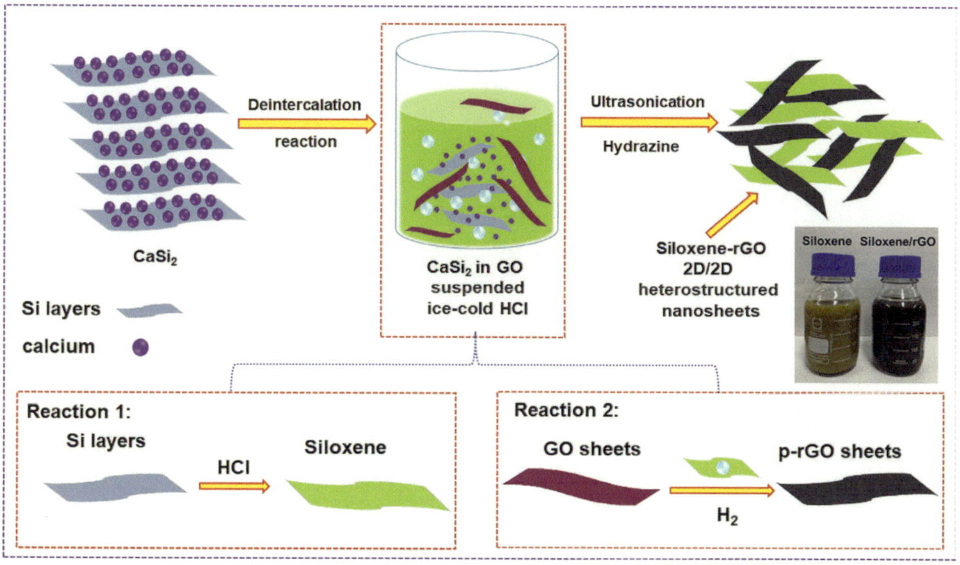

Figure 1.
Schematic illustration of the synthesis of siloxene-rGO heterostructures. Reproduced with permission: Copyright 2021, Wiley-VCH.

2D/2D heterostructures were re-dispersed in an aqueous solution containing hydrazine hydrate and subject to ultrasound irradiation for the reduction of GO into rGO sheets. The bare siloxene sheets were prepared by the same reaction without the addition of GO, and the bare rGO was prepared by sonochemical reaction, as mentioned in our earlier works [32, 37].

Figure 2(a) presents the X-ray diffraction (XRD) pattern of the siloxene-rGO heterostructures in comparison with the siloxene and GO sheets. The presence of peaks at the lower diffraction angles, i.e., 11.3° and 10.4° confirmed the formation of siloxene and GO sheets by the topochemical de-intercalation reaction and modified Hummer's method, respectively [10, 27, 38]. The XRD pattern of the siloxene-rGO powders shows the presence of a broad peak at 23.3°, indicating the formation of the siloxene-rGO hybrids via random interstratification. Further, the absence of the peaks observed in bare siloxene and GO sheets confirmed the removal of oxygen-bonded functional groups from their basal planes and edges, respectively. **Figure 2(b)** shows the Fourier transform-infra red (FT-IR) spectrum of siloxene, GO, and siloxene-rGO powders. The presence of oxygenated functional groups like carbonyl, carboxyl, and epoxide, in addition to the C-C vibration in the GO, closely matched well with the literature [39]. The presence of Si-H, Si-O-Si, Si-Si, Si-OH, and $O\text{-}Si_2 = Si\text{-}H$ bands in the FT-IR spectrum of siloxene powders confirms their Kautsky-type nature [32, 40]. The FT-IR spectrum of siloxene-rGO powders shows the presence of Si-O-Si, and $O\text{-}Si_2 = Si\text{-}H$ bands raised from the siloxene and C-C and carbonyl groups of rGO, suggesting the formation of siloxene-rGO powers via the removal of surface oxygen groups present in the individual sheets. The comparative X-ray photoelectron spectroscopic (XPS) Si 2p spectra of the $CaSi_2$, siloxene, and siloxene-rGO powders (**Figure 2(c)**) displayed pronounced changes in their surface states. The Si-Si states observed at 99.5 eV disappeared after hydrazine reaction, whereas the Si-O-Si states (102.5 eV) were preserved. This suggested that the hydrazine treatment not only reduces the GO sheets but also reacts with siloxene sheets which needed to be studied in detail

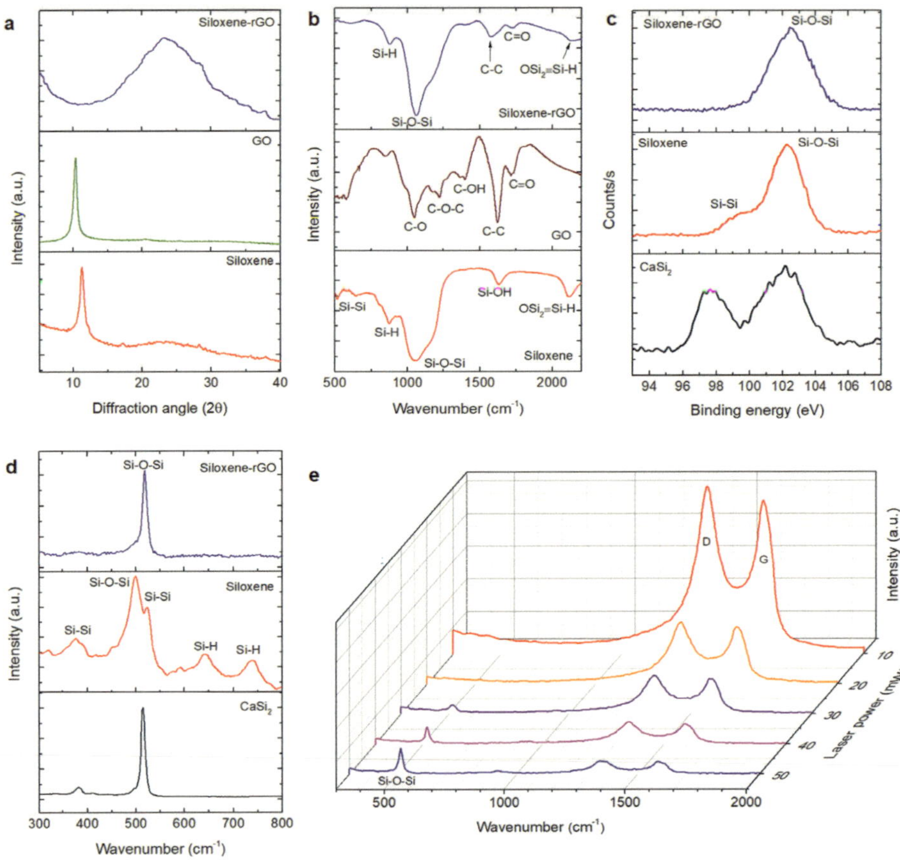

Figure 2.
(a) XRD patterns of siloxene-, GO-, and siloxene-rGO powders. (b) FT-IR spectra of siloxene-, GO-, and siloxene-rGO powders. (c) Si 2p core-level spectra of $CaSi_2$-, siloxene-, and siloxene-rGO powders. (d) Laser Raman spectra of $CaSi_2$-, siloxene-, and siloxene-rGO at lower wavenumber ranges. (e) Laser Raman spectrum of siloxene-rGO powders with respect to laser power levels. Reproduced with permission: Copyright 2021, Wiley-VCH.

in the near future. The XPS C 1 s analysis of rGO (control sample) and the siloxene-rGO heterostructures (data not shown) revealed that the hydrazine treatment under ultrasound irradiation effectively reduces the GO into rGO [10]. **Figure 2(d)** shows the laser Raman spectra of the $CaSi_2$ precursor, siloxene sheets, and siloxene-rGO heterostructures. All of the Raman signatures of the siloxene powders such as from Si-H, and were Si-Si disappeared in the Raman spectrum of siloxene-rGO powders whereas Si-O-Si is retained with slight peak shift toward a higher wavenumber [41]. The Raman spectra of the siloxene-rGO powders (**Figure 2(e)**) displayed the presence of G band (1582 cm^{-1}) and D band (1350 cm^{-1}) of rGO sheets at all ranges of input laser power. At the same time, the bands raised from siloxene sheets (Si-O-Si) were seen only at the high powers [10]. **Figure 3(a-d)** depicts the morphological analysis of the siloxene-rGO heterostructures using high resolution transmission electron microscopic (HR-TEM) analysis with various magnification levels. It confirms the presence of siloxene sheets on rGO's surface. The observation of hexagonal diffraction spots in the selected area electron diffraction (SAED) pattern (inset of **Figure 3(d)**) was raised from the rGO sheets present in the heterostructures. The elemental mapping

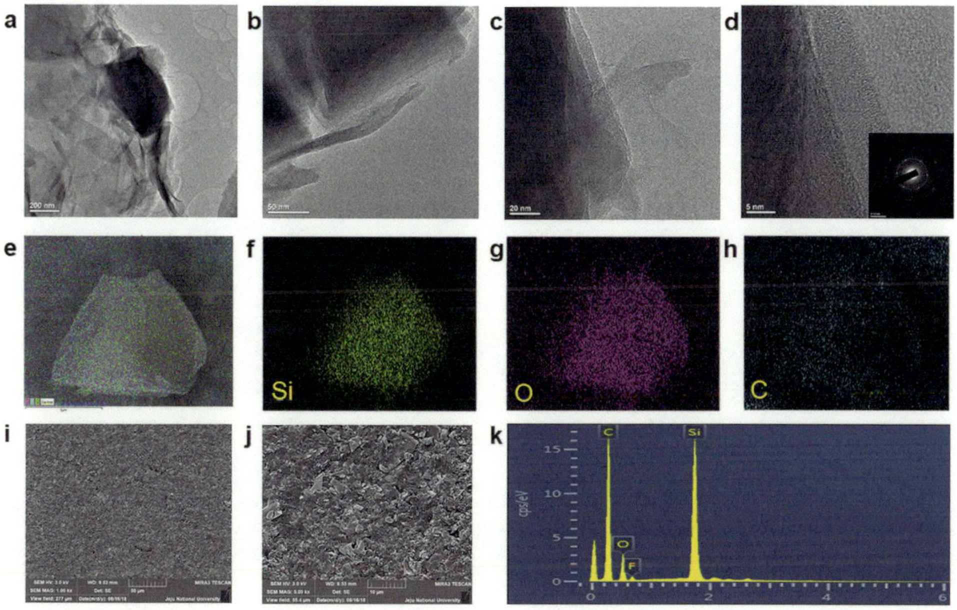

Figure 3.
(a-d) HR-TEM images of siloxene-rGO powders under different magnifications; the inset in (d) shows their SAED pattern. (e-h) Elemental maps of siloxene-rGO powders with (e) overlay micrograph, and (f-h) represents the Si, O and C maps. (i-j) field emission-scanning electron microscopic (FE-SEM) images of roll pressed siloxene-rGO SC electrodes. (k) energy-dispersive X-ray spectroscopy (EDX) spectrum of roll pressed siloxene-rGO SC electrode. Reproduced with permission: Copyright 2021, Wiley-VCH.

analysis given in **Figure 3(e-h)** confirms the homogeneous distribution of Si element, O element, and C element in the prepared siloxene-rGO heterostructures in which both sheets are randomly stacked with each other. Further, the FE-SEM micrograph of the siloxene-rGO powders in the form of SC electrode fabricated using the Doctor blade method (**Figure 3(i-j)**) indicated the lateral orientation of the individual sheets. The elemental spectrum of the fabricated SC electrode (**Figure 3(k)**) confirms that they are made of Si, O, and C elements.

In order to study the supercapacitive properties of the siloxene-rGO heterostructures, a symmetric type SC in the form of coin-cell configuration was constructed using 1 M $TEABF_4$ electrolyte. The standard practice methods viz. cyclic voltammetry (CV), charge-discharge (CD) analysis, and electrochemical impedance spectroscopy (EIS) were used to determine the performance metrics of the siloxene-rGO SC in comparison with the bare siloxene and rGO SCs. **Figure 4(a)** compares the CV profiles of the fabricated SCs measured at 50 mV s^{-1}. It showed that all the SCs operate over 3.0 V without any evolution; however, the current ranges of the siloxene-rGO SC are higher than the bare siloxene and rGO SCs, indicating the better energy storage performance of siloxene-rGO SC. The scan rate effect on the CV profiles of the siloxene-rGO SC is shown in **Figure 4(b-c)**, which evidences no significant deviation in the rectangular shape of their profiles, suggesting good conductivity of the heterostructures. **Figure 4(d)** compares the device capacitance of the fabricated SCs at different sweep rates. It confirmed the superior charge-storage performance of the siloxene-rGO SC with a high device capacitance of 56.17 F g^{-1} compared to the rGO SSC (37.55 F g^{-1}) and siloxene SSC (4.89 F g^{-1}) at 5 mV s^{-1}. Further, **Figure 4(e)** confirms

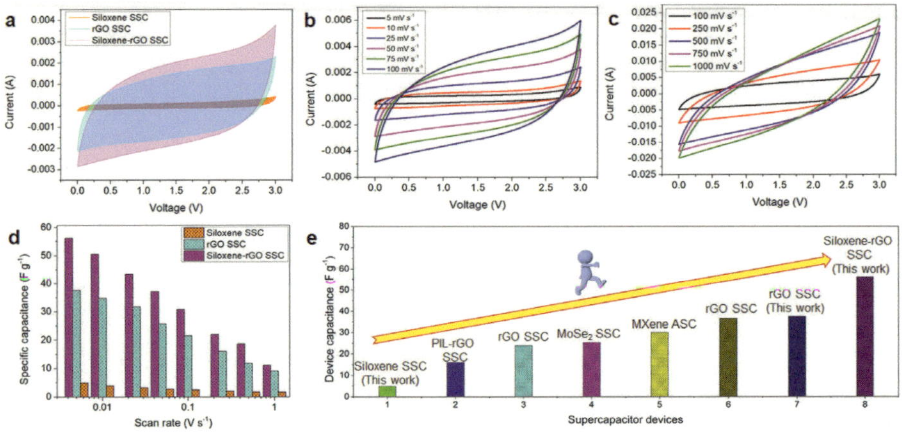

Figure 4.
(a) CV profiles of siloxene, rGO, and siloxene-rGO SCs recorded at 50 mV s^{-1} over 3.0 V. (b-c) CVs of siloxene-rGO SC recorded at different scan rates. (d) Effect of applied scan rates on the device capacitance of the fabricated SCs. (e) Comparison of the siloxene-rGO SC's capacitance with recently reported ones. Reproduced with permission: Copyright 2021, Wiley-VCH.

the exceptional performance of the siloxene-rGO SC over the reported 2D materials/their hybrid-based SCs [10]. **Figure 5(a)** displays the comparative CD profiles of the fabricated SCs measured by using 1 mA current. It showed that the ultra-fast charge-discharge properties of siloxene SC due to the ion-intercalation capacitance, whereas rGO SC possesses better charge-storage properties compared to siloxene SC due to the EDLC nature of rGO [39]. However, the siloxene-rGO SC possesses higher charging and discharging time compared to the others indicating the role of the synergetic effects of ion-intercalation and EDL properties of the heterostructures [10]. The CD profiles given in **Figure 5(b)** revealed symmetric triangular-shaped profiles of the siloxene-rGO SC that were retained even after an increased in applied current ranges (**Figure 5(c)**). The plot of device capacitance with respect to the applied current of the fabricated SCs is compared in **Figure 5(d)**. It showed that the siloxene-rGO SC possesses a capacitance of 44.63 F g$^{-1,}$ which is higher than the siloxene SC (1.79 F g^{-1}) and rGO SC (30.46 F g^{-1}) from the CD profiles (@1 mA). Further, **Figure 5(e)** demonstrates the excellent power ratings of the siloxene-rGO SC to handle various currents. **Figure 5(f)** highlights the better long-term electrochemical stability of the siloxene-rGO SC over 10,000 cycles of continuous CD profiles with excellent capacitance retention of 112%. The initial increase in capacitance during the initial cycles (upto 2500 cycles) is due to the electro-activation process involved in the siloxene-rGO electrodes [10].

Figure 5(g) provides the Ragone plot of the fabricated SCs for understanding the improvements in electrochemical performances of the heterostructures compared to individual counterparts. The siloxene-rGO SC possesses an extra-ordinary specific energy of 55.79 Wh kg^{-1}, which is better than the rGO SC (38.07 Wh kg^{-1}) and siloxene SC (2.23 Wh kg^{-1}) at a specific power of 1500 W kg^{-1}. The specific energy of the siloxene-rGO SC varies from 55.7 to 16.2 Wh kg^{-1}, when their specific power increases from 1.5 to 15 kW kg^{-1}. From **Figure 5(g)**, it is evident that the performances of the siloxene-rGO SC are higher compared to the state-of-the-art SCs reported in literature [42–44]. Strikingly, the specific energy of the siloxene-rGO SC surpasses the commercial SCs and reaches the level of lead–acid batteries [32, 45]. The excellent metrics of siloxene-rGO SC originate from the structural benefit of 2D/2D heterostructure and

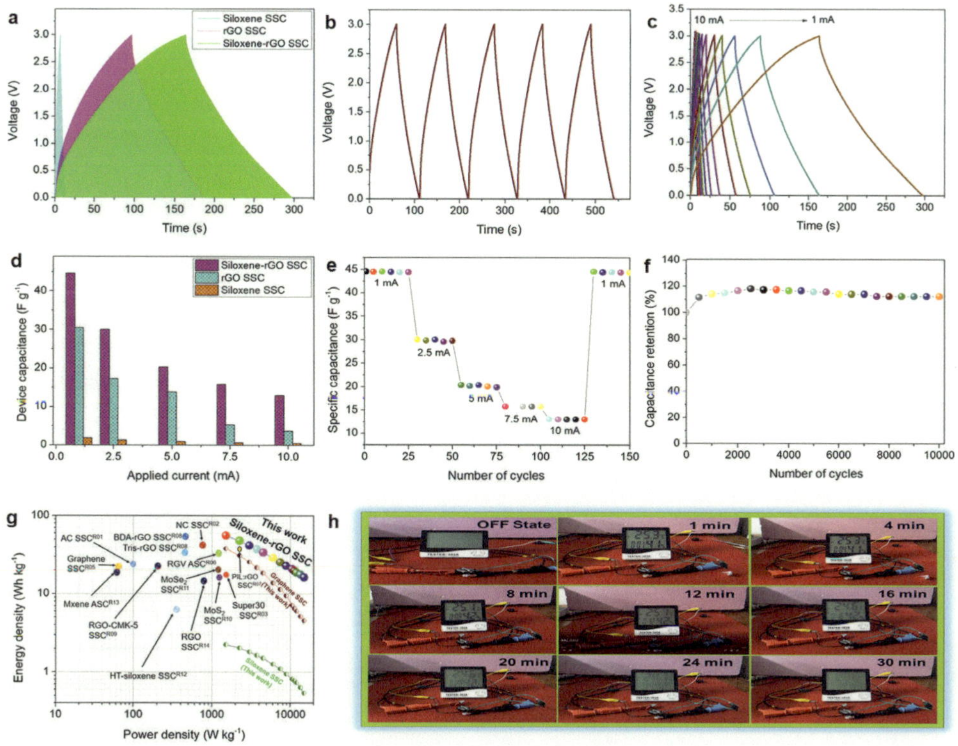

Figure 5.
(a) The CD profiles of siloxene, rGO, and siloxene-rGO SCs recorded using 1 mA. (b) Continuous CD profiles of siloxene-rGO SCs measured using 2 mA. (c) CD profiles of siloxene-rGO SCs measured using different applied current ranges. (d) Effect of applied current on the capacitance of the fabricated SCs. (e) Rate capability of siloxene-rGO SC. (f) Long-term stability of siloxene-rGO SC. (g) Ragone chart of siloxene-rGO SCs in comparison with reported SCs. (h) Practical applications of fully charged siloxene-rGO SC used to power a multifunctional electronic display (MFED) over 30 minutes. The references in **Figure 5(g)** *can be seen from ref [10]. Reproduced with permission: Copyright 2021, Wiley-VCH.*

differential electrochemical properties of the individual sheets of the heterostructure [46, 47]. **Figure 5(h)** portrays the applications of the siloxene-rGO SC as a sustainable energy storage and delivery system. The siloxene-rGO SC was charged upto 3.0 V, and the stored charges were used for continuous powering of a multifunctional electronic display (MFED) over 30 min in which all of their functionalities (humidity, temperature, and clock) can be monitored. **Figure 6** shows the supercapacitive properties of the siloxene-rGO SC measured over a broad temperature range (−15 to 80°C). **Figure 6(a)** depicts the CV profiles of the siloxene-rGO SC measured at 100 mV s^{-1} under various temperature (−15 to 80°C). It shows the change in the current ranges without distorting the rectangular shaped CV profiles. As seen from **Figure 6(b)**, the siloxene-rGO SC possess a capacitance of 9.23 and 16.45 F g^{-1} at a low temperature of −15 and 5°C whereas they tend to rise from 16.45 to 34.86 F g^{-1} when the temperature reaches to 80°C. The temperature-dependent Nyquist plot of siloxene-rGO SC (**Figure 6(c)**) revealed that the high charge-transfer resistance (R_{ct}) at low temperatures since ion diffusion kinetics are limited at extremely low temperatures, and they tend to be minimal at high temperatures [48]. The Warburg line shifts toward the imaginary axis with the change in temperature (−15 to 80°C), suggesting increased capacitive nature for the siloxene-rGO SC device at high temperatures. The Bode modulus plot of the

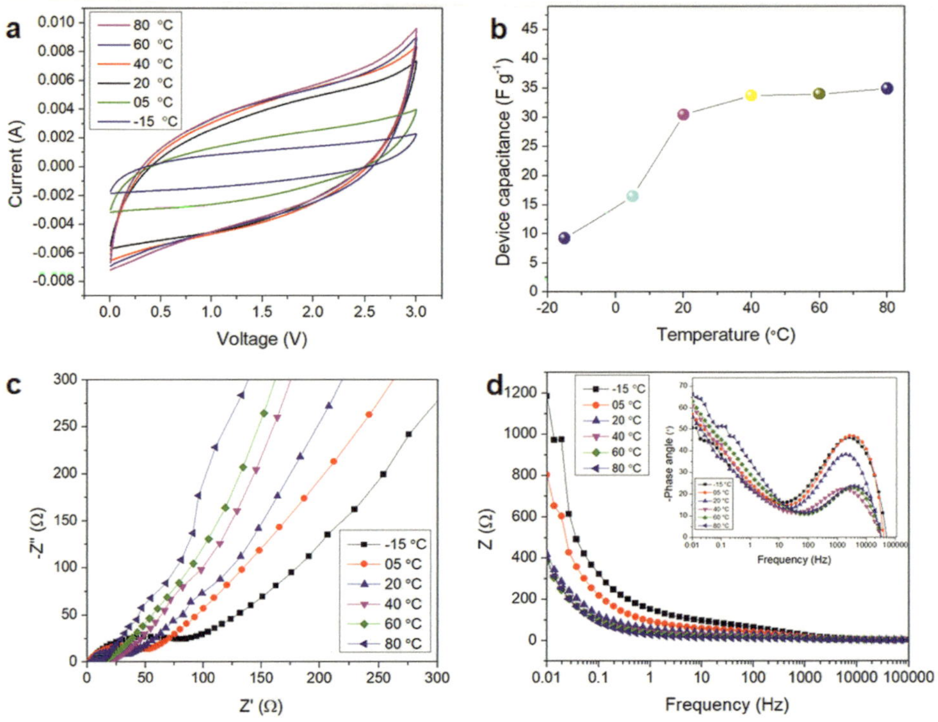

Figure 6.
(a) CV profiles of siloxene-rGO SC over broad temperature range. (b) Effect of temperature on the siloxene-rGO SC's capacitance. (c) Nyquist plot of siloxene-rGO SC, and (d) bode modulus plot of siloxene-rGO SC. The inset in figure (d) presents bode phase angle plot of siloxene-rGO SC. Reproduced with permission: Copyright 2021, Wiley-VCH.

siloxene-rGO SC (**Figure 6(d)**) shows the low impedance at high temperatures and vice versa. The Bode phase angle plot of the siloxene-rGO SC (inset in **Figure 6(d)**) revealed the phase angle at 0.01 Hz shifts from −50.57° (−15°C) to −66.13° (80°C), highlighting their better capacitive properties at high temperatures [49].

The practical applicability of the siloxene-rGO SCs as an ideal EES system for the recovery of regenerative braking energy (RRBE) process in EVs is demonstrated experimentally, as seen in **Figure 7**. The main condition which makes the SC an ideal candidate for regenerative braking systems (RBSs) is their power density greater than15 000 W kg^{-1}, an essential power target suggested by the Partnership for New Generation of Vehicles (PNGVs) [50]. Based on the schematic representation of the RBS in EVs (**Figure 7(a)**), a prototype RBS system was designed (**Figure 7(d)**) comprising a wheel with two circular brakes (controlled by piston) and electric motors (act as a generator to convert kinetic energy into electrical energy) to which the siloxene-rGO SC was connected. The amount of charge stored with the various levels of braking condition is provided in **Figure 7(b)**. It shows at the initial 25 s when the RBS is in the rest state, whereas when the brake is applied, the siloxene-rGO SC was charged to 0.8 V (@ 30 s) and 1.2 V (@ 90 s) and with the rapid braking, they charged up to 2.2 V (**Figure 7(c)**), confirming their capability toward an efficient EES for EVs [51]. **Figure 7(e-f)** shows the effective use of siloxene-rGO SC combined with the RBS system to light up the braking lights. In addition to the RBS, two parallelly connected siloxene-rGO SC was charged via an EV power station using solar cells

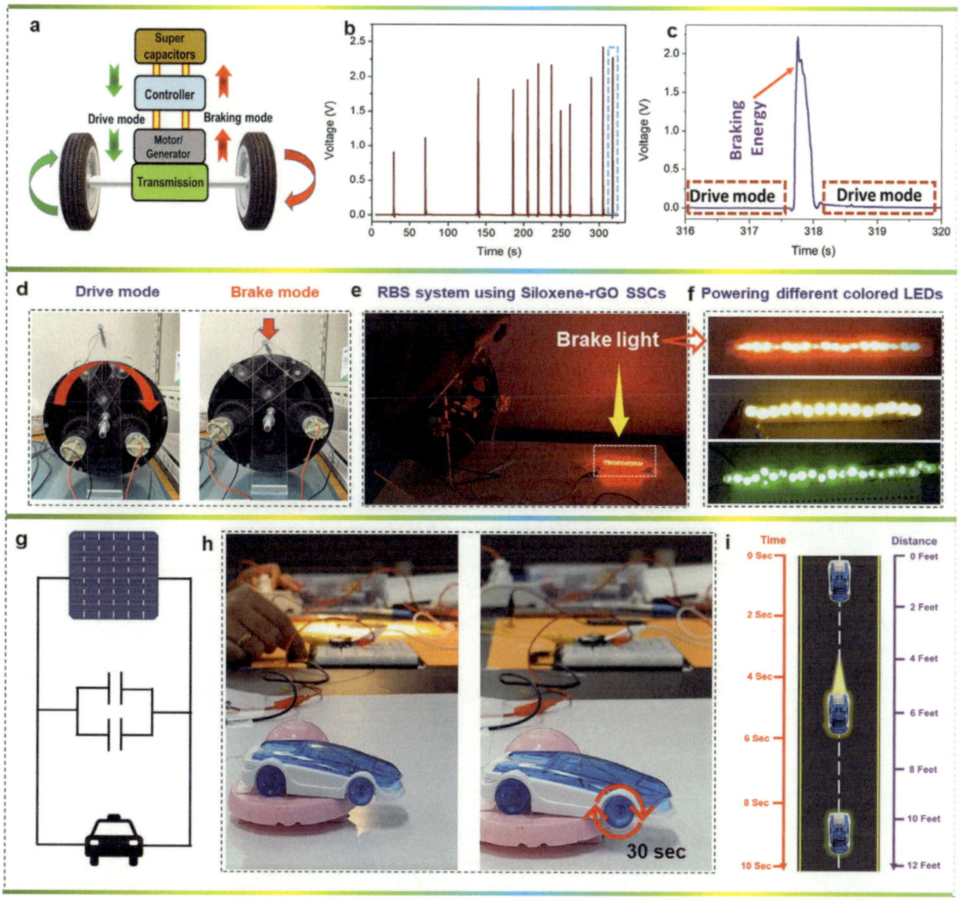

Figure 7.
(a) Principle of the RRBE process in EVs. (b) Voltage vs. time profile of siloxene-rGO SC under driving and braking conditions. (c) Magnified portion of the region marked in (c) that indicated the SCs was charged up to 2.2 V under high braking conditions. (d) Digital photograph of the RRBE prototype. (e-f) Practical utilization of the siloxene-rGO SC-based RBS system for powering up braking light with different LEDs. (g) EVs charging station, (h) digital photograph of charging using EV station (left) and siloxene-rGO SC's ability to drive an EV for 30 seconds in stand-by mode (right). (i) On-road test of EV driven by siloxene-rGO SC over 11 feet within 9 seconds. Reproduced with permission: Copyright 2021, Wiley-VCH.

and used for driving a toy EV over ≈30 s (**Figure 7(g-h)**) at stationary conditions. Similarly, the on-road test (**Figure 7i**) shows the ability of the siloxene-rGO SCs to drive the EV over 11 feet within 9 s. Overall, the experimental results demonstrate that the siloxene-rGO SSC can function as a powerful and reliable power source for portable electronics and EVs.

3. Conclusions

This chapter demonstrated the synthesis of siloxene-rGO heterostructures and their applications for supercapacitors in EVs. The 2D/2D heterostructures were prepared via multi-step chemical reactions (viz (i) topochemical de-intercalation and (ii) the ultrasound-assisted reduction reaction., respectively. Further, the physico-chemical characterizations suggested that heterostructures were formed via random interstification kinetics. The siloxene-rGO SCs using ionic liquid hold high device capacitance, specific energy, and specific power with long cycle life. Their performance metrics were higher compared to the individual rGO, siloxene sheets, and many of the reported 2D materials-based SCs. More importantly, the fabricated siloxene-rGO SCs possess excellent temperature tolerance, thereby ensuring their utilization at various temperatures. Owing to their high power density, the experimental results demonstrated the ability of the siloxene-rGO SCs to store energy generated from the regenerative braking in EV prototypes, thus, ensuring their potential applications as an EES system in next-generation EVs.

Acknowledgements

The Basic Science Research Program supported this research through the National Research Foundation of Korea (NRF) grant funded by the Korean Government (MSIT) (2021R1A4A2000934 and 2023R1A2C3004336).

Conflict of interest

The authors declare no conflict of interest.

Author details

Karthikeyan Krishnamoorthy[1,2,3*], Parthiban Pazhamalai[2,3], Rajavarman Swaminathan[2] and Sang-Jae Kim[2,3,4*]

1 Department of Physics, School of Advanced Sciences, Vellore Institute of Technology, Vellore, India

2 Nanomaterials and System Laboratory, Faculty of Applied Energy System, Major of Mechatronics Engineering, Jeju National University, Jeju, South Korea

3 Research Institute of New Energy Industry (RINEI), Jeju National University, Jeju, South Korea

4 Nanomaterials and System Laboratory, Major of Mechanical System Engineering, College of Engineering, Jeju National University, Jeju, South Korea

*Address all correspondence to: karthi.nanotech@gmail.com and kimsangj@jejunu.ac.kr

References

[1] Muratori M, Alexander M, Arent D, Bazilian M, Cazzola P, Dede EM, et al. The rise of electric vehicles—2020 status and future expectations. Prog. Energy. 2021;**3**:022002. [Internet]. Available from: https://iopscience.iop.org/article/10.1088/2516-1083/abe0ad

[2] Sudjoko C, Sasongko NA, Utami I, Maghfuri A. Utilization of electric vehicles as an energy alternative to reduce carbon emissions. IOP Conference Series: Earth and Environmental Science. 2021;**926**:012094. [Internet]. Available from: https://iopscience.iop.org/article/10.1088/1755-1315/926/1/012094

[3] Gupta A, Kumar N. Energy regeneration in electric vehicles with wind turbine and modified alternator. Materials Today: Proceedings. 2021;**47**:3380-3386 [Internet]. Available from: https://linkinghub.elsevier.com/retrieve/pii/S2214785321050239

[4] Hunt JD, Jurasz J, Zakeri B, Nascimento A, Cross S, ten Caten CS, et al. Electric truck hydropower, a flexible solution to hydropower in mountainous regions. Energy. 2022;**248**:123495. [Internet]. Available from: https://linkinghub.elsevier.com/retrieve/pii/S036054422200398X

[5] Diahovchenko I, Petrichenko L, Borzenkov I, Kolcun M. Application of photovoltaic panels in electric vehicles to enhance the range. Heliyon. 2022;**8**:e12425. [Internet]. Available from: https://linkinghub.elsevier.com/retrieve/pii/S2405844022037136

[6] Castelvecchi D. Electric cars and batteries: How will the world produce enough? Nature. 2021;**596**:336-339. [Internet]. Available from: https://www.nature.com/articles/d41586-021-02222-1

[7] Chen W, Liang J, Yang Z, Li G. A review of lithium-ion battery for electric vehicle applications and beyond. Energy Procedia. 2019;**158**:4363-4368. [Internet]. Available from: https://linkinghub.elsevier.com/retrieve/pii/S1876610219308215

[8] Manoharan S, Krishnamoorthy K, Sathyaseelan A, Kim S-J. High-power graphene supercapacitors for the effective storage of regenerative energy during the braking and deceleration process in electric vehicles. Materials Chemistry Frontiers. 2021;**5**:6200-6211. [Internet]. Available from: http://xlink.rsc.org/?DOI=D1QM00465D

[9] Lin C-L, Hung H-C, Li J-C. Active control of regenerative brake for electric vehicles. Actuators. 2018;7:84. [Internet]. Available from: http://www.mdpi.com/2076-0825/7/4/84

[10] Krishnamoorthy K, Pazhamalai P, Mariappan VK, Manoharan S, Kesavan D, Kim S. Two-dimensional Siloxene–graphene Heterostructure-based high-performance Supercapacitor for capturing regenerative braking energy in electric vehicles. Advanced Functional Materials. 2021;**31**:2008422. [Internet]. Available from: https://onlinelibrary.wiley.com/doi/10.1002/adfm.202008422

[11] Geraee S, Mohammadbagherpoor H, Shafiei M, Valizadeh M, Montazeri F, Feyzi MR. Regenerative braking of electric vehicle using a modified direct torque control and adaptive control theory. Computers and Electrical Engineering. 2018;**69**:85-97. [Internet]. Available from: https://

linkinghub.elsevier.com/retrieve/pii/S004579061733272X

[12] Yibre M, Abido MA. Supercapacitors for wind power application. In: 2013 International Conference on Renewable Energy Research and Applications (ICRERA). IEEE. 2013. pp. 418-422. [Internet]. DOI: 10.1109/ICRERA.2013.6749791

[13] Mariappan VK, Krishnamoorthy K, Pazhamalai P, Sahoo S, Kesavan D, Kim SJ. Two dimensional famatinite sheets decorated on reduced graphene oxide: A novel electrode for high performance supercapacitors. Journal of Power Sources. 2019;**433**:126648. [Internet]. Available from: https://linkinghub.elsevier.com/retrieve/pii/S0378775319306172

[14] Wang J, Malgras V, Sugahara Y, Yamauchi Y. Electrochemical energy storage performance of 2D nanoarchitectured hybrid materials. Nature Communications. 2021;**12**:3563. [Internet]. Available from: https://www.nature.com/articles/s41467-021-23819-0

[15] Lee CW, Suh JM, Choi S, Jun SE, Lee TH, Yang JW, et al. Surface-tailored graphene channels. npj 2D Materials and Applications. 2021;5:39. [Internet]. Available from: https://www.nature.com/articles/s41699-021-00223-2

[16] Su J, He W, Li X-M, Sun L, Wang H-Y, Lan Y-Q, et al. High electrical conductivity in a 2D MOF with intrinsic Superprotonic conduction and interfacial pseudo-capacitance. Matter. 2020;**2**:711-722. [Internet]. Available from: https://linkinghub.elsevier.com/retrieve/pii/S2590238519304096

[17] Acerce M, Voiry D, Chhowalla M. Metallic 1T phase MoS2 nanosheets as supercapacitor electrode materials. Nature Nanotechnology. 2015;**10**:313-318. [Internet]. Available from: http://www.nature.com/articles/nnano.2015.40

[18] Pazhamalai P, Krishnamoorthy K, Sahoo S, Kim SJ. Two-dimensional molybdenum diselenide nanosheets as a novel electrode material for symmetric supercapacitors using organic electrolyte. Electrochimica Acta. 2019;**295**:591-598. [Internet]. Available from: https://linkinghub.elsevier.com/retrieve/pii/S0013468618324538

[19] Ratha S, Rout CS. Supercapacitor electrodes based on layered tungsten disulfide-reduced graphene oxide hybrids synthesized by a facile hydrothermal method. ACS Applied Materials & Interfaces. 2013;**5**:11427-11433. [Internet]. Available from: http://pubs.acs.org/doi/10.1021/am403663f

[20] Krishnamoorthy K, Pazhamalai P, Sahoo S, Kim SJ. Titanium carbide sheet based high performance wire type solid state supercapacitors. Journal of Materials Chemistry A. 2017;**5**:5726-5736. [Internet]. Available from: http://xlink.rsc.org/?DOI=C6TA11198J

[21] Simon P, Gogotsi Y. Perspectives for electrochemical capacitors and related devices. Nature Materials. 2020;**19**:1151-1163

[22] Jiang Q, Lei Y, Liang H, Xi K, Xia C, Alshareef HN. Review of MXene electrochemical microsupercapacitors. Energy Storage Materials. 2020;**27**:78-95

[23] Mariappan VK, Krishnamoorthy K, Pazhamalai P, Natarajan S, Sahoo S, Nardekar SS, et al. Antimonene dendritic nanostructures: Dual-functional material for high-performance energy storage and harvesting devices. Nano Energy. 2020;**77**:105248. [Internet]. Available from: https://linkinghub.elsevier.com/retrieve/pii/S2211285520308260

[24] Zhan C, Zhang P, Dai S, Jiang DE. Boron Supercapacitors. ACS Energy Letters. 2016;**1**:1241-1246. [Internet]. Available from: http://pubs.acs.org/doi/10.1021/acsenergylett.6b00483

[25] Tao Y, Huang T, Ding C, Yu F, Tan D, Wang F, et al. Few-layer phosphorene: An emerging electrode material for electrochemical energy storage. Applied Materials Today. 2019;**15**:18-33

[26] Krishnamoorthy K, Pazhamalai P, Mariappan VK, Nardekar SS, Sahoo S, Kim SJ. Probing the energy conversion process in piezoelectric-driven electrochemical self-charging supercapacitor power cell using piezoelectrochemical spectroscopy. Nature Communications. 2020;**11**:2351. [Internet]. Available from: http://www.nature.com/articles/s41467-020-15808-6

[27] Pazhamalai P, Krishnamoorthy K, Sahoo S, Mariappan VK, Kim SJ. Understanding the thermal treatment effect of two-dimensional Siloxene sheets and the origin of superior electrochemical energy storage performances. ACS Applied Materials & Interfaces. 2019;**11**:624-633. [Internet]. Available from: http://pubs.acs.org/doi/10.1021/acsami.8b15323

[28] Wang X, Kajiyama S, Iinuma H, Hosono E, Oro S, Moriguchi I, et al. Pseudocapacitance of MXene nanosheets for high-power sodium-ion hybrid capacitors. Nature Communications. 2015;**6**:6544. [Internet] Available from: http://www.nature.com/doifinder/10.1038/ncomms7544

[29] Tang Y, Chen T, Yu S, Qiao Y, Mu S, Zhang S, et al. A highly electronic conductive cobalt nickel sulphide dendrite/quasi-spherical nanocomposite for a supercapacitor electrode with ultrahigh areal specific capacitance. Journal of Power Sources. 2015;**295**:314-322

[30] Zhang LL, Zhao X, Stoller MD, Zhu Y, Ji H, Murali S, et al. Highly conductive and porous activated reduced graphene oxide films for high-power supercapacitors. Nano Letters. 2012;**12**:1806-1812. [Internet]. Available from: http://pubs.acs.org/doi/10.1021/nl203903z

[31] Chhowalla M, Shin HS, Eda G, Li L-J, Loh KP, Zhang H. The chemistry of two-dimensional layered transition metal dichalcogenide nanosheets. Nature Chemistry. 2013;**5**:263-275. [Internet]. Available from: http://www.nature.com/articles/nchem.1589

[32] Krishnamoorthy K, Pazhamalai P, Kim SJ. Two-dimensional siloxene nanosheets: Novel high-performance supercapacitor electrode materials. Energy. Environmental Sciences. 2018;**11**:1595-1602. [Internet]. Available from: http://xlink.rsc.org/?DOI=C8EE00160J

[33] Zhang X, Hou L, Ciesiclski A, Samorì P. 2D materials beyond graphene for high-performance energy storage applications. Advanced Energy Materials. 2016;**6**:1600671. [Internet]. Available from: https://onlinelibrary.wiley.com/doi/full/10.1002/aenm.201600671

[34] Wang J, Nie P, Ding B, Dong S, Hao X, Dou H, et al. Biomass derived carbon for energy storage devices. Journal of Materials Chemistry A. 2017;**5**:2411-2428. [Internet]. Available from: http://xlink.rsc.org/?DOI=C6TA08742F

[35] Jeon H, Jeong JM, Kang HG, Kim HJ, Park J, Kim DH, et al. Scalable water-based production of highly conductive 2D Nanosheets with ultrahigh volumetric

capacitance and rate capability. Advanced Energy Materials. 2018;**8**:1-11

[36] Panda PK, Grigoriev A, Mishra YK, Ahuja R. Progress in supercapacitors: Roles of two dimensional nanotubular materials. Nanoscale Advances. 2020;**2**:70-108

[37] Krishnamoorthy K, Pazhamalai P, Sahoo S, Lim JH, Choi KH, Kim SJ. A high-energy aqueous sodium-ion capacitor with nickel Hexacyanoferrate and graphene electrodes. ChemElectroChem. 2017;**4**:3302-3308. [Internet]. Available from: http://doi.wiley.com/10.1002/celc.201700690

[38] Yu J, Zhou J, Yao P, Xie H, Zhang M, Ji M, et al. Antimonene engineered highly deformable freestanding electrode with extraordinarily improved energy storage performance. Advanced Energy Materials. 2019;**9**:1902462

[39] Krishnamoorthy K, Manoharan S, Mariappan VK, Pazhamalai P, Kim S-J. Decoupling mechano- and electrochemical gating: A direct visualization for piezo-ionic propelled proton tunneling in self-charging supercapacitors. Journal of Materials Chemistry A. 2022;**10**:7818-7829. [Internet]. Available from: http://pubs.rsc.org/en/Content/ArticleLanding/2022/TA/D1TA10254K

[40] Yamanaka S, Matsu-ura H, Ishikawa M. New deintercalation reaction of calcium from calcium disilicide. Synthesis of layered polysilane. Materials Research Bulletin. 1996;**31**:307-316. [Internet]. Available from: http://linkinghub.elsevier.com/retrieve/pii/0025540895001956

[41] Oughaddou H, Enriquez H, Tchalala MR, Yildirim H, Mayne AJ, Bendounan A, et al. Silicene, a promising new 2D material. Progress in Surface Science [Internet]. 2015;**90**:46-83. DOI: 10.1016/j.progsurf.2014.12.003

[42] Pohlmann S, Olyschläger T, Goodrich P, Alvarez Vicente J, Jacquemin J, Balducci A. Azepanium-based ionic liquids as green electrolytes for high voltage supercapacitors. Journal of Power Sources. 2015;**273**:931-936

[43] Song B, Zhao J, Wang M, Mullavey J, Zhu Y, Geng Z, et al. Systematic study on structural and electronic properties of diamine/triamine functionalized graphene networks for supercapacitor application. Nano Energy. 2017;**31**:183-193. [Internet]. Available from: https://linkinghub.elsevier.com/retrieve/pii/S2211285516304797

[44] Lei Z, Liu Z, Wang H, Sun X, Lu L, Zhao XS. A high-energy-density supercapacitor with graphene–CMK-5 as the electrode and ionic liquid as the electrolyte. Journal of Materials Chemistry A. 2013;**1**:2313. [Internet]. Available from: http://xlink.rsc.org/?DOI=c2ta01040b

[45] Ji H, Zhao X, Qiao Z, Jung J, Zhu Y, Lu Y, et al. Capacitance of carbon-based electrical double-layer capacitors. Nature Communications. 2014;**5**:3317

[46] Bissett MA, Kinloch IA, Dryfe RAW. Characterization of MoS2-graphene composites for high-performance coin cell Supercapacitors. ACS Applied Materials & Interfaces. 2015;**7**:17388-17398. [Internet] [cited 2017 Dec 9]. Available from: http://pubs.acs.org/doi/10.1021/acsami.5b04672

[47] Velický M, Toth PS. From two-dimensional materials to their heterostructures: An electrochemist's perspective. Applied Materials Today. 2017;**8**:68-103

[48] Lu C, Chen X. All-temperature flexible Supercapacitors enabled by

Antifreezing and thermally stable hydrogel electrolyte. Nano Letters. 2020;**20**:1907-1914

[49] Krishnamoorthy K, Pazhamalai P, Kim SJ. Ruthenium sulfide nanoparticles as a new pseudocapacitive material for supercapacitor. Electrochimica Acta. 2017;**227**:85-94. [Internet] [cited 2017 Aug 11]. Available from: https://linkinghub.elsevier.com/retrieve/pii/S001346861632730X

[50] Wang D-WW, Li F, Liu M, Lu GQ, Cheng H-MM. 3D aperiodic hierarchical porous graphitic carbon material for high-rate electrochemical capacitive energy storage. Angewandte Chemie International Edition. 2008;**47**:373-376. [Internet]. Available from: http://doi.wiley.com/10.1002/anie.200702721

[51] Teymourfar R, Asaei B, Iman-Eini H, Nejati FR. Stationary super-capacitor energy storage system to save regenerative braking energy in a metro line. Energy Conversion and Management. 2012;**56**:206-214

Chapter 6

Graphene Nanosheets for Fuel Cell Application

Barsha Das, Sagnik Das, Soumyabrata Tewary, Sujoy Bose, Sandip Ghosh and Avijit Ghosh

Abstract

The unique chemical and physical properties of graphene enable it to be an alternative and better component in fuel cells as seen in recent studies. Fuel cells provide constant energy supply without any degradation of power over time with a continuous supply of fuel and show very little amount of emission of hazardous pollutants. Fuel cells can be a potent replacement for conventional petroleum energy sources. The limitations of producing efficient fuel cells are mainly high cost of materials, short lived membrane due to corrosion, cells being heavier in weight, efficient polymer electrolyte membrane (PEM), less efficient electrocatalysts, and slow reaction rates. To minimize all these limitations, graphene, a 2D allotrope of carbon, can be used due to its excellent properties; i.e., high surface area, high conductivity, high proton permeability, better electro catalytic performance, lower cost, greater corrosion resistivity and high bonding energy to hydrogen. In this chapter, the various components of fuel cells are discussed where graphene nanosheets and its derivatives (Graphene oxides and heteroatom doped graphene) are used to improve the fuel cells performance efficient.

Keywords: graphene nanosheet, fuel cells, electrocatalyst, ORR, corrosion resistance, proton conductivity, power density

1. Introduction

The use of global energy sources is extensively increasing due to the rapid growth and development of industrialization. The existing energy sources like coal, oil and natural gases are limited in nature, and energy production from these sources has a huge negative effect on the environment [1]. Energy production from renewable sources has much less impact on the environment but they also have some drawbacks such as low efficiency, higher cost [2]. The drawback led to the study of highly efficient energy sources, that is, fuel cell [3].

The significance of fuel cell is analyzed by the rapid rise of application worldwide in the past decade. A fuel cell generates electricity through an electrochemical reaction. It converts the chemical energy of a fuel and an oxidizing agent into electricity through redox reactions [2]. As the chemical energy of fuels converts into electricity by chemical reaction, it has a much higher efficiency than combustion cells along with low emission of pollutants; the by-products being only water [4]. Fuel cells need not to be charged, they can produce electricity under continuous fuel supply. If the waste heat can be utilized properly, up to 85% efficiency can be obtained from fuel cells [5].

According to the types of electrolytes used, fuel cells are classified into phosphoric acid fuel cell (PAFC), polymer electrolyte membrane fuel cell (PEMFC), alkaline fuel cell (AFC), molten carbonate fuel cell (MCFC), and solid oxide fuel cell (SOFC). Apart from these, there are different kinds of fuel cells, like direct methanol fuel cell (DMFC), and direct formic acid fuel cell (DFAFC) [1, 4]. Microbial fuel cells (MFC) can produce energy by microbial oxidation and can be used for wastewater treatment in a cost-effective and sustainable way [6].

A fuel cell mainly consists of bipolar plates (BPs), a cathode, an anode, and an electrolyte. The main reactions in a fuel cell include fuel oxidation on the anode and oxygen reduction on the cathode [1, 7]. The material selection for fuel cell components faces challenges including electrochemical performance, efficiency, economy, and durability [4]. The materials mostly used for bipolar plates include non-porous graphitic carbon and conducting polymers. This shows corrosion after a longer period of use and its mechanical strength is also poor. Metal and metal oxides (Pt, RuO_2, and IrO_2) are used as electrocatalysts. But the cost and availability of the materials limit their application. Also, RuO_2 oxidizes to RuO_4 under high anodic potential, which reduces the active surface area, therefore, decreasing efficiency. Self-poisoning created by the high adsorption rate of CO on the Pt surface reduces its efficiency as an electrocatalyst. Pt is also not a good electrocatalyst for organic fuels [3]. Electrodes like Pt/C after a certain period of use show corrosion of carbon and aggregation of Pt which decreases the efficiency of the fuel cell [8]. Sulfonated tetrafluoroethylene-based fluoropolymer-copolymer (Nafion), dow membranes, per-fluorinated ionomers are mainly used as membrane or electrolyte in the fuel cell [2]. These polymers can work at an optimum rate under a certain temperature range. Changes in temperature can destabilize the system and decrease efficiency [9]. Their performance is also affected by hydrated conditions and fuel crossover issues [4].

A graphene nanosheet is a one-atom thick layer of sp^2 hybridized carbon arranged in a honeycomb structure. After its discovery in 2004, graphene is under research for different applications due to their excellent properties such as very high theoretical surface area (2600 $m^2\ g^{-1}$), high mechanical strength (100–200 times stronger than steel), high thermal stability and excellent electrical conductivity (10^6 S cm^{-1}), and excellent corrosion resistance [7]. Recent advancements in the use of graphene nanosheets for fuel cell applications have shown some promising results to overcome the problems discussed above related to the practical use of fuel cells [4]. As a result of the expansion of research, various derivatives of graphene are reported to be produced, like graphene oxide, multilayered graphene, porous graphene, porous graphene oxide, and impurity-doped graphene [10].

Graphite is the most used material for bipolar plates but its mechanical strength is poor and it comes with a higher fabrication cost. Metallic plates have much higher electrical conductivity but show higher corrosion rates in an acidic environment. Graphene with an anti-corrosive nature has been used as a protective layer for bipolar plates [11]. Graphene-coated copper plates showed 3 times more charge transfer resistance than uncoated copper after 720 hours of immersion in 0.5 mol L^{-1} H_2SO_4 solution [12]. A sharp increase in electrical conduction was observed using graphene-coated Ni foam due to numerous micropores in the three-dimensional Ni foam. By this coating, the corrosion resistance of the bipolar plate is increased by almost 2 times [13]. Graphene-based materials can also be used as fillers to build up highly conductive and durable polymeric bipolar plates. A highly filled graphene/poly benzoxazine composite with 60 wt% graphene content, exhibiting high flexural modulus (18 GPa), flexural strength (42 MPa), thermal conductivity (8.0 W m^{-1} K^{-1}), electrical conductivity (357 S cm^{-1}), and low water absorption (0.06% at 24 hours immersion) [4, 14].

One of the graphene derivatives, graphene oxide (GO), contains oxygen functional groups with very high surface area, which results in a higher transfer rate of protons through channels and holds the water for better water uptake. Various properties of graphene nanosheets such as electrical insulation, gas permeability, and hydrophilicity make it a promising material for composite membranes in PEMFCs [2]. Graphene composite membranes can work in a much higher temperature range than currently available membranes used. F-GO/Nafion shows almost 4 times higher proton conductivity than unmodified Nafion at 30% relative humidity and 120°C [15]. By incorporating 0.5 wt% ionic liquid polymer-modified graphene sheets, the ionic conductivity (7.5×10^{-3} S cm^{-1} at 160°C) of sulfonated polyimide membrane was improved by about 2.6 times [16]. Graphene is flexible for working with proton-conducting groups like sulfonic acid to further facilitate proton transport [17]. A critical drawback of the Nafion membrane is the fuel crossover, especially when using methanol. Graphene can act as a barrier layer to stop fuel gas permeability without compromising the proton exchange action which increases overall cell performance [18].

Graphene and graphene-based materials like graphene oxides show structural properties which help them to possess carbonyl, hydroxy, and epoxy on their basal plane and the carboxylic groups at the edges. These oxygen-containing groups enhance graphene to make a desired structure to exhibit high conductivity and stability [8]. The PEMFC with Pt-Co/N-doped graphene cathode showed a four times higher maximum power density (805 mW cm^{-2} at 60°C) than that of a commercial Pt/C cathode. Graphene-supported single-atom catalysts (SACs) show high oxygen reduction reaction activity and long-term stability in alkaline or acidic media [4]. Pt/$NrEGO_2$-CB_3 maintains 86.35% of the initial electrochemical surface area after 30,000 cycles compared to 2% of the electrochemical surface area of Pt/C [19]. atomically dispersed Ru on N-doped graphene exhibited higher ORR activity, better durability, and tolerance toward methanol and CO poisoning than commercial Pt/C catalyst in 0.1 $molL^{-1}$ $HClO_4$ [20]. The MFCs suffer from low power density and poor energy conversion efficiency due to the slow transfer rate of an electron from electron donors to electrodes producing electricity. Graphene having high conductivity and high biocompatibility can be used in MFCs to enhance electron transfer rate by increasing surface area and reducing the internal resistance [21].

In this chapter, the use of graphene nanosheets and its derivatives are described below for the various fuel cell components, such as, electrocatalyst, membrane, and bipolar plate.

2. Graphene nanosheets: synthesis, characterization, and properties

2.1 Synthesis of graphene nanosheets

The first graphene nanosheets were synthesized by extracting mono-layer sheets from three-dimensional graphite with Scotch tape using a technique called micromechanical cleavage [3]. Since then, a wide range of synthesis processes has been developed and have been broadly divided into two types; the bottom-up method and top-down method. The bottom-up method includes mainly chemical vapor deposition (CVD) on catalytically active metals, epitaxial growth on single crystal SiC, and arc discharge approach [3, 22]. The top-down method includes mechanical exfoliation, chemical exfoliation, ball milling, ultrasonic treatment, electrochemical exfoliation, liquid-phase exfoliation, and high-shear mixing [4, 23]. In chemical vapor deposition method, graphene nanosheets are prepared on transition metal substrates such as Ni, Pd, Ru, Ir, and Cu and the thickness is dependent on the carbon solubility in these metals and their cooling rate [22]. The transfer process of the CVD is time-consuming and contaminated whereas electrochemical-assisted de-lamination of the CVD provides a fast, environmentally friendly, and more controllable approach to graphene production [4]. Different substrates are used in CVD for graphene film growth; they include Ni, Fe, Cu, and stainless steel [24]. High-quality graphene can be produced on SiC wafers under high temperature [22]. This process is limited by the high cost of the single crystal SiC and, due to the wafer size, the graphene nanosheets are produced at a limited size [4]. In the arc discharge approach, carbon nanosheets are prepared by evaporating carbon sources at high temperatures [3]. In the top-down synthesis process, a single layer of graphene is obtained by mechanical exfoliation [4]. In chemical exfoliation, graphene nanosheets are obtained with fewer defects by a stable alkali metal salt intercalation compound which is industrially applicable [22]. In liquid phase exfoliation, graphene is derived from organic solvent with certain surface tension provided by external forces like ball milling, ultrasonic treatment, and shear mixing [3]. It can be noted that graphene can be synthesized by using various methods as reported in the literature. Moreover, its characterization is to be supported for the successful synthesis. The various characterization methods of graphene nanosheets are described in the subsequent section.

2.2 Characterization of graphene nanosheets

Characterization of graphene is an important aspect to understand its physical and chemical properties. The development of graphene and its derivatives were found through different physical and chemical methods. Characterization includes study of graphene morphology, properties, defects, and layered structure. The nanomaterials are dispersed in the matrix or substrate material through exfoliation and then studied [24, 25].

There are different characterization methods includes Fourier Transform Infrared Spectroscopy (FTIR), Solid-state nuclear magnetic resonance (SS-NMR), X-ray Photoelectron Spectroscopy (XPS), Raman Spectroscopy, Thermo-Gravimetric Analysis (TGA), and different kind of microscopic technique including Transmission Electron Microscopy (TEM), Scanning Electron Microscopy (SEM), and Atomic Force Microscopy (AFM) [26].

FTIR can be used to detect the functional groups of the specimen sample. In case of graphene oxides presence of different oxygen containing functional groups can be identified by the vibrational peaks obtained. The chemical changes in transition from edges to the inner region of GO flakes can be seen in the FTIR images [25, 26]. The electrical and mechanical properties can be varied with the presence of functional group. The SS-NMR is used to determine the properties of functional group doped graphene nanosheets. XPS provides some advantages over FTIR and SS-NMR in giving both basic information and quantification about the sample. It also shows the percentage of different oxygen containing functional group on the graphene surface

[26]. Raman spectroscopy has an advantage in studying graphene with polar impurities. Three peaks (D, G, and 2D) obtained in Raman spectroscopy explains disorder in sp^2 hybridization, lattice vibration, and stacking degree of graphene respectively. TGA is used to study the graphene characteristics and its stability under different temperatures [24, 26]. SEM, TEM, and AFM are mainly used to analysis the surface morphology of the graphene nanosheets and its derivatives. It helps in the study of size of the nano particles and layered structure of graphene [24–26].

It can be seen that there are many methods to characterize the graphene nanosheet for its wide range of applications. Based on requirements, some of the methods are to be used to characterize graphene, which helps to choose the perfect graphene nanoparticles which provide expected results. Below the sections will discuss the various properties of graphene and its derivatives.

2.3 Properties of graphene nanosheets and derivatives of graphene

2.3.1 Properties of graphene nanosheets

Graphene and graphene derivatives exhibit exceptional physical and chemical properties, such as, large specific area, high electrical and thermal conductivity, greater durability, and chemical strength. These excellent properties make graphene one of the most appropriate materials for fuel cell components [4].

Graphene has a theoretical surface area of 2630 $m^2\ g^{-1}$ which is very much higher than graphite powders. It has a mechanical strength of young modulus of 1100GPa. It shows stable properties even at a temperature higher than 1000°C [27]. It shows high electrical conductivity at an order of 106 S cm^{-1} and electrical mobility of 200,000 $cm^2\ V^{-1}\ s^{-1}$at a carrier density of ~1012 cm^2. It has a breaking stress of 42 N m^{-1} and excellent thermal conductivity (~5000 W $m^{-1}\ K^{-1}$) [28]. The approximate thickness of a single layer of graphene is 0.345 nm. It is a transparent monolayer structure which shows properties of light absorption. In addition, the atomic C-C bonds at 284 eV and the wavenumber at 1550 cm^{-1} have been recorded with the narrow scan of X-ray photoelectron spectroscopy (XPS) and Fourier transformation infrared (FTIR) spectra, respectively [10]. Good thermal conductivity helps in an application that requires strict heat management and the reactions exhibiting a strong exothermicity or endothermicity [29]. Other than graphene nanosheets, there are other derivatives, namely, graphene oxide and heteroatom doped graphene, which are greatly applicable in fuel cell research.

2.3.2 Properties of graphene oxide

Graphene oxide (GO) nanosheet is a single-layered structure usually produced by chemical oxidation of graphite. Oxygen functional groups like hydroxyl groups and epoxy groups are found on the basal planes, other groups like carboxy, carbonyl, phenol etc. were mostly formed at the sheet edges. These oxygen containing groups and defects change its electrical and thermal properties. GO is strongly hydrophilic and disperses uniformly in water, which makes it an excellent electrode material [30]. GO has a high proton conductivity (1.1×10^{-5} - 2.8×10^{-3} S cm^{-1}) making it an excellent choice for polymer electrolyte membrane materials in PEMFC [4]. GO is unlikely to work at high temperatures and hence, limiting its use at low temperature applications [31]. However, the electrical conductivity of GO is lower than graphene due to present of functional group on the surface.

2.3.3 Properties of heteroatom doped graphene

The properties of graphene can be tuned and modified as per the requirements by doping heteroatoms into the basal planes and reactive edges. It is reported that the doping increases the active site and enhances electrocatalytic activity of graphene significantly [32]. Nitrogen is the most popular dopant used owing to its chemical stability against chemical oxidation [32]. However, carrier mobility and conductivity of N-doped graphene are slightly lower than pristine graphene due to hinder effect. On the other hand, the Boron-doped and thiophene rich S-doped graphene show higher conductivity and charge transfer ability. Hydrogenation of graphene can control its semiconducting properties [33].

From the above it can be understand that graphene nanosheet has the several properties which can be used to enhance the fuel cell electrochemical activity as compared to commercial Pt/CB electrocatalyst. Subsequent section will discuss the application of graphene nanosheet for the various component in PEM fuel cell and microbial fuel cell.

3. Application of graphene nanosheets in fuel cells

3.1 Polymer electrolyte membrane fuel cell (PEMFC)

3.1.1 Electrocatalyst support

The effectiveness of the fuel cell depends on the Hydrogen evolution reaction (HER), Oxygen evolution reaction (OER), and Oxygen reduction reaction (ORR) at the electrode. These reactions proceed at a very slower rate, decreasing the efficiency of the overall fuel cell. Metals and metal oxides, such as, Pt, RuO_2, and IrO_2 are the most popular electrocatalysts used to increase the rate of the reactions [4, 8]. Although these electrocatalysts show promising results, various reasons limit their wide implementation, including very high cost, oxidation of RuO_2, CO poisoning of Pt, limited abundance etc. [34].

Graphene or its derivatives are an excellent alternative to these conventional electrocatalysts due to its comparatively lower cost and unique properties. Graphene is a zero-band gap semiconductor which decreases its catalytic activity, but doping of graphene with different heteroatoms varies its electronic properties and increases active sites and enhances its electrocatalytic activity significantly [32]. The morphology of a graphene-based electrocatalyst can significantly affect the different electrocatalytic reactions in PEMFC. The different morphological structures of graphene nano sheets can be obtained by distorting graphene nanosheets either by topological defects, heteroatom doping, or through creation of active edges [34]. Conductive graphene derivatives are used to facilitate electron transfer rates. Many graphene composites show positive results solving many critical problems associated with PEMFC [4]. By topological defect engineering, the number of active sites can be increased in graphene nanosheets. Interfacial effect engineering in graphene-based electrocatalysts can be used to produce low-cost, high-performance OWS (Overall water splitting) electrocatalyst [34]. Heteroatom doped graphene can be used as a support for single metal atoms, which increases active site area largely and as a result it shows high ORR activity and selectivity for the 4-electron route. It also provides long term stability and a greater charge distribution to the system [4, 20].

In **Table 1**, the performances of graphene based electrocatalyst are presented. A heteroatom doped graphene molecule Ru-N/G catalyst was synthesized for the application in PEMFC as an electrocatalyst. The ORR electrocatalytic performance of the Ru-N/G catalystwas calculated by rotating disk electrode (RDE) measurements which was performed in acidic electrolyte (0.1 M $HClO_4$) at a rotation speed of 1600 rpm. Based on different annealing temperatures in the presence of NH3, these products are sub-categorized into Ru-N/G-550, Ru-N/G-650, Ru-N/G-750, and Ru-N/G-800. The products obtained at lower annealing temperature show lower current density and at higher temp (~800°C) Ru shows agglomeration. So, the optimum electrocatalytic behavior is shown by Ru-N/G-750. Mass activity of the ORR for Ru-N/G-750 is 7.5 times higher than that of the commercial Pt/C, thus increasing metal utilization and decreasing the cost for electrocatalyst. Durability of the electrocatalyst is notably increased. After 10,000 cycles, the catalyst holds up to 93% of initial saturated current [20]. Reduced graphene oxide supported platinum ($Pt/NrEGO_2\text{-}CB_3$) is also a graphene derivative which showed promising results under considerable testing. It shows very high charge retention, a higher energy density than commercial Pt/C, and lower mass transfer resistance [19]. In a recent study, it was reported that Co_3O_4 nanocrystals on reduced modified graphene oxide (rmGO) can be used as a catalyst

Graphene based material	Properties & Performance	References
Ru-N/G-750	• ORR activity is 7.5 times higher than commercial Pt/C catalyst system. • Lower manufacturing cost. • Exceptional high durability. • Exhibits onset potential at 0.89 V and half wave potential of 0.75 V.	[20]
$Pt/NrEGO_2\text{-}CB_3$	• Maintains high effective surface area value after 30,000 cycles ($63.83\ cm^2\ g^{-1}_{pt}$) • Offers smaller amount of resistance in compared to commercial electrocatalysts. • Shows high power density of 0.411 W cm^{-1}. • Lower level of Pt agglomeration under stressed condition.	[19]
Co_3O_4/N-rmGO	• Low cost of the catalyst • High durability • High ORR activity	[35]
Ni/rGO	• High mass activity (1600 mA mg^{-1}) • High stability (maintains 1012 mA mg^{-1} after 1000 cycles)	[4]
Fe-Co-N/rGO-700	• Shows Over potential of 0.215 V for HER at the current density of 10 mA cm^{-2}. • Shows over potential of 0.308 V for OER at current density of 10 mA cm^{-2}. • Shows Cell potential of 1.5 V	[36]
Pt/FG	• ESA after 1000th scan is 53.33 $m^2\ g^{-1}_{Pt}$ • Forward peak current density after 1000th scan is 42.06 mA cm^{-2}	[37]

Table 1.
Summary of properties and performances of recent-past graphene based Electrocatalyst.

for Oxygen reduction reaction (ORR). The composite material showed similar properties to commercially available Pt/C composites but its durability is very high in adverse conditions and manufacturing cost is much lower [35]. By highly dispersing ultrafine (~2.3 nm) Ni nanoparticles on reduced graphene oxide (rGO), the composite showed high mass activity of 1600 mA m g^{-1} and high stability [4]. The nitrogen doped Fe-Co/rGO electrocatalyst improved electrocatalytic performance significantly (for HER and OER) in alkaline solution. The over potentials is reported of 0.215 V for HER and 0.308 V for ORR at 10 mA $cm^{-2.}$ The developed catalyst shows a high durability (after 45 h) [36]. Using Pt/functionalized graphene (Pt/FG) in comparison to Pt/C gives a higher amount of electrochemical active surface area (ECSA) and forward peak current density (mA cm^{-2}) [37].

3.1.2 Reinforcement in the polymer electrolyte membrane

Polymer electrolyte membranes (PEM) play a vital role in the PEMFC fuel cell system [1]. Commercially used PEMs include Nafion, Flemion, and Aciplex, which are mainly based on Perflurosulfonic acid (PFSA). Though they all show promising results, different conditions restrict its widespread application [17]. The most common problems related to PEMs are low rate of performance in highly humid conditions and high working temperature; fuel crossover issues; high cost; critical synthesis methods; and water balancing issues [2, 17, 38]. The development of a PEM with adequate conductivity operating under high temperature range and low humid conditions has become a primary objective of the fuel cell technology [9]. Impregnation of graphene in PEM enhances reaction rate, thus, improving the efficiency and the durability of fuel cell devices [1, 4]. The application of graphene

Polymers	Graphene material	Graphene content	Advancements	References
Sulfonated poly (ether etherketone)	Sulfonated reduced graphene oxide	1.0 wt%	• Very less amount of gas crossover. • Increase in open circuit voltage. • High proton conductivity.	[17]
Sulfonated polyamide	Ionic liquid polymer modified graphene sheets	10 wt%	• Ionic conductivity, tensile strength and mechanical properties were improved.	[16]
		0.5 wt%	2.6 times better proton conductivity than unaltered membrane.	
Nafion	Polyoxometalate coupled graphene oxide	1 wt%	• Increase in proton conductivity. • Reduction in ohmic resistance. • Water retention. • Maximum power density of 841 mW cm^{-2}	[39]
Polybenzimi dazole (PBI)	3-amino propyl-triethoxysilane ionic liquid with functionalized Graphite oxide	5 wt%	Improves proton conductivity and ionic conductivity significantly.	[40]

Table 2.
Graphene nanosheet/polymer composites for PEMFCs.

and its 2D derivatives has proved to be profitable due to its properties like large electrical conductivity, high charge carrier rates, vast surface area, and a lower production cost.

A PEM must offer high proton permeability, greater electrochemical and thermal stability, high water uptake rates, low permeability to reactant species, high mechanical stability, and lower cost [2]. **Table 2** shows the comparison in the properties of graphene/polymer nanocomposites for fuel cell applications.

Sulfonated reduced graphene oxide (SRGO) as a conductive layer incorporated with sulfonated poly (ether ether ketone) (SPEEK). Among many tryouts, SPEEK/SRGO-1.0 membrane shows excellent proton conductivity also under low relative humidity (RH). It shows high proton conductivity of 8.6 mScm-1 at 80°C/50% RH, which is 3 times greater than commercially available SPEEK membranes. Also, a higher power output of 705 mW cm^{-2} is generated compared to 636 mW cm^{-2} for non-modified membrane [17]. Another study shows graphene modified protic ionic liquid-based membrane which improves ionic conductivity, proton conductivity, tensile strength and mechanical properties of fuel cell remarkably at a lower production cost [16]. Modification of Nafions with Polyoxometalate coupled with graphene oxide show higher amounts of proton conductivity and power density compared to nafion alone. It also increases its efficiency at working for a longer period [39].

It is also seen that modified graphene oxide material and incorporated it with the polybenzimidazole membrane material and improved its protonic and ionic conductivity notably [40].

3.1.3 Graphene nanosheet reinforced carbon-polymer composite bipolar plate

Bipolar plate (BP) is one of the important parts in PEMFCs. Its function includes provide support to the cell, distribute fuel and oxidant to the electrode surface connecting, collect current from the cells. Therefore, it should have high electrical conductivity, good mechanical strength and low gas permeability [7, 41]. Both metals (Ti, Ni etc.) and non-metals (graphite) are used as the material for bipolar plates in PEMFCs. Graphite is one of the important carbon materials now a days attracted as the reinforcement in the carbon-polymer composite for bipolar plates in PEMFCs. Because, graphene has high corrosion resistance, low specific gravity, and high electrical conductivity (10^6 Scm^{-1}) [4, 11]. However, carbon-polymer bipolar plate suffers lack of mechanical strength. This can be overcome using metallic bipolar plate. Metallic plate has high electrical conductivity and high mechanical strength, but it suffers due to corrosion problem [27]. By incorporating graphene as a supporting material for bipolar plates in both metallic and non-metallic bipolar plates, researchers found some significant improvement in the results. 1% graphene reinforcement increased by around 6% and 35% in cases of in-plane and through plane electrical conductivities of graphitic bipolar plate composites with phenyl-formaldehyde resin (RPFR) as binder [7, 11].

It has been presented in **Table 3**, Polyphenylene sulfide (PPS) as a matrix, compositing it with graphene nanoplatelets (G-NPs), making a new bipolar plate material which showed excellent properties under different GNP loading according to different working conditions [13]. The second material presented is by using graphene 0.2 parts per hundred parts of resin (phr) in graphitic bipolar plate composite with vinyl ester resin (VER) as binder shows promising thermal conductivity and flexure strength values are found [7]. It has been proposed graphene-Poly benzoxazine composites as bipolar plate material based on the promising experimental results. 60 wt% of graphene loading showed optimum results in electrical conductivity and

Bipolar plate material	Electrical Conductivity ($S\ cm^{-1}$)	Flexure strength (MPa)	Corrosion current ($\mu A\ cm^{-2}$)	References
Graphene/Carbon Fiber/Carbon Black/Natural Graphite/RPFR 1/5/5/64/25 vol%	435.31^{i} 130.17^{t}	57.28	NA	[7, 11]
G-NP/PPS (60 wt% G-NP)	1.7	66	NA	[13]
Graphene/VER/Natural Graphite 0.2phr/70/30 wt%	286.4^{i}	49.2	NA	[7]
G-NP/ Poly benzoxazine (60 wt% G-NP)	357	41.7	NA	[14]
G-NP/Poly Propylene (7.5 wt% G-NP)	5.28×10^{-11}	NA	NA	[27]
GO/PPY/SS304 (1 mg mL^{-1} GO)	NA	NA	0.059	[27]
RMGO/Ti	NA	NA	<1	[41]

t = through-plane; i = in-plane; NA = not available.

Table 3.
Graphene nanosheet reinforced bipolar plates.

flexure strength [14]. Graphene nano particles (7.5 wt%) and polypropylene composite showed electrical conductivity of 5.28×10^{-11} S cm^{-1}. Graphene oxide polypyrole (PPY) composite showed 0.059 μA cm-2 of corrosion current, which is very low compared to commercial metallic BPs [27]. Reduced modified graphene oxide (RMGO), titanium composite bipolar plate showed corrosion current of <10–6 A cm^{-2}, and interfacial contact resistance (ICR) is as low as 4 mΩ cm^2 [41].

3.2 Microbial fuel cell

Microbial fuel cells are cells which convert chemical energy into electrical energy. The cell performs critical fuel oxidation reaction on the cathode while oxygen reduction reactions on the anode. Among many metals used, graphene nanosheets exhibit better electrolytic performance with increases in power density as well as current density [42]. In **Table 4**, the recent studies that apply material doped graphene nanosheets to increase the electrocatalytic activities reduction reaction (ORR) catalytic activities in the cathode as well as in the anode are tabulated.

To enhance the cathode performance, studies show cobalt nanoparticles embedded in nitrogen doped carbon nanosheets coupled with graphene show excellent ORR activity. The compound pyrolyzed at 800°C optimizes the ORR activity. Moreover, it also provides long term stability [43]. The B-doped graphene quantum dots also serve as a durable ORR cathode catalyst which gives a power density 1.5 times the Pt/C cathode [44]. Iron (III) oxide and boron nitride-doped reduced graphene oxide nanosheets enhance the cathode performance determined by impedance analysis which is conducted over the range of 1 MHz to 100 MHz at 10 mV [42]. Polyaniline graphene nanosheets improve electricity generation capacity with a maximum power density of 99 mW m^{-2}, promoting electrolytic performance [45].

To enhance the anode performance, nitrogen doped graphene is used to exhibit maximum electrooxidation current [46]. Titanium suboxide with graphene as anode provides a steady voltage of 980 mV, highly enhancing charge transfer efficiency [47].

Fuel cell component	Doping compound with graphene	Maximum power density (mW m^{-2})	Open circuit potential (mV)	Electro chemical active surface area (m^2 g^{-1})	Reference
Cathode	Cobalt nanoparticles embedded in N doped carbon shell with graphene	611 ± 9	NA	NA	[43]
	B-doped graphene	703.55	NA	NA	[44]
	Iron (III) oxide and boron nitride-doped reduced graphene oxide nanosheets	2066 ± 15	663 ± 4	73.1638	[42]
	Polyaniline-graphene nanosheets	Improved from 0.85 to 99	580	NA	[45]
Anode	Nitrogen doped graphene sheet	1008	NA	NA	[46]
	Titanium suboxides with graphene	2073	980	26.5	[47]

Table 4.
Applications of graphene composites in MFCs.

4. Conclusions

This chapter summarized the progress currently done by implementing graphene-based material on fuel cell components, such as, electrode, electrocatalysts, and bipolar plates. It is observed that the graphene, owing to its excellent properties including thermal, chemical, and electro-chemical properties showed significant improvement in each of the application.

Ruthenium (Ru) is a commercially used electrocatalysts in PEMFCs. Its ORR activity is improved by 7.5 times when graphene was used to make Ru-N/G-750 composite. It also showed higher durability and lower synthesis cost. On the other hand, Pt/NrEGO$_2$-CB$_3$ composite electrocatalyst shows higher value of ECSA of 63.83 cm^2 $g^{-1}{}_{pt}$ even after 30,000 cycles of use and provides lower resistance and higher power density. Several modified electrocatalyst discussed in this chapter which includes Co_3O_4/N-rmGO, Ni/rGO, Fe-Co-N/rGO-700, and Pt/FG. All these catalysts show enhanced electrochemical activities in PEMFCs. The HAD and ORR activity of the graphene nanosheet supported electrocatalyst improved significantly.

It is also added that the GO reinforced SPEEK membrane nanocomposite improved the properties significantly. This nanocomposite membrane reduces the gas crossover, increased open circuit voltage, and enhanced proton conductivity. Moreover, it can be observed that the sulfonated polyamide, Nafion, and PBI all this membrane modified by using graphene showed promising results also under adverse conditions.

The Graphene reinforced carbon-polymer bipolar plate showed excellent electrical conductivity of 435.32 S cm^{-1}, 130.17 S cm^{-1} respectively for in-plane and through-plane conditions and showed flexure strength of 57.28 Mpa. The use of graphene nanosheet in the various component of PEMFC is shows the improved in the overall performance significantly including power density of a single cell and/or fuel cell stack. Therefore, the use of graphene nanosheet in other fuel cell also attracted to the researchers and scientist.

In MFCs both cathode and anode materials are enhanced by using graphene nanosheets and its derivative. Using Cobalt nanoparticles embedded in N-doped carbon shell with graphene as cathode in MFC gives a high-power density of 611 mW m^{-2}. Iron (III) oxide and boron nitride doped rGO cathode in MFC showed higher power density of ~2066 mW m^{-2}, open circuit potential of ~663 mV and ECSA of 73.1638 m^2 g^{-1}. Other different electrode compound such as B-doped graphene, Polyaniline graphene nanosheets, N-doped graphene, and Titanium suboxides with graphene shows greater performances in MFCs. From this study, it can be attributed that the graphene nanosheets and its derivatives provide a new approach to fabricating high performance fuel cell components including excellent electrocatalyst in the simulated environment.

Acknowledgements

The authors gratefully acknowledge the contribution of Indian Institute of Chemical Engineers for guidance and support in writing this book chapter.

Appendices and nomenclature

AFC	Alkaline fuel cell
AFM	Atomic force microscopy
BPs	Bipolar Plates
CVD	Chemical vapor deposition
DFAFC	Direct formic acid fuel cell
DMFC	Direct methanol fuel cell
ECSA	Electro chemical active surface area
FTIR	Fourier Transform Infrared Spectroscopy
F-GO	Functionalized graphene
G-NPs	Graphene nano platelets
GO	Graphene Oxide
HER	Hydrogen evolution reaction
MFC	Microbial fuel cell
MCFC	Molten carbonate fuel cell
OWS	Overall water splitting
OER	Oxygen evolution reaction
ORR	Oxygen reduction reaction
Phr	Parts per hundred parts of resin
PFSA	Perflurosulfonic acid
PAFC	Phosphoric acid fuel cell
PBI	Polybenzimidazole
PEM	Polymer electrolyte membrane
PEMFC	Polymer electrolyte membrane fuel cell
PPS	Polyphenylene sulfide
PPY	Polypyrole
rEGO	reduced Exfoliated Graphene Oxide

rGO	reduced Graphene Oxide
rmGO	reduced modified Graphene Oxide
RH	Relative humidity
RDE	Rotating Disk electrode
RPFR	phenyl-formaldehyde resin
SEM	Scanning Electron Microscopy
SACs	Single atom catalysts
SOFC	Solid oxide fuel cell
SS-NMR	Solid State Nuclear Magnetic Resonance
SPEEK	Sulfonated poly (ether ether ketone)
SRGO	Sulfonated reduced Graphene Oxide
TGA	Thermo Gravimetric Analysis
TEM	Transmission Electron Microscopy
VER	Vinyl ester resin
XPS	X-ray Photoelectron Spectroscopy

Author details

Barsha Das[1], Sagnik Das[1], Soumyabrata Tewary[1], Sujoy Bose[2], Sandip Ghosh[2] and Avijit Ghosh[1,2*]

1 Department of Chemical Engineering, Heritage Institute of Technology, Anandapur, West Bengal, India

2 Indian Institute of Chemical Engineers, Jadavpur University Campus, Kolkata, India

*Address all correspondence to: avijit.ghosh@heritageit.edu

References

[1] Iqbal MZ, Siddique S, Khan A, Haider SS, Khalid M. Recent developments in graphene based novel structures for efficient and durable fuel cells. Materials Research Bulletin. 2020;**122**:110674. DOI: 10.1016/j.materresbull.2019.110674

[2] Farooqui UR, Ahmad AL, Hamid N. A graphene oxide: A promising membrane material for fuel cells. Renewable and Sustainable Energy Reviews. 2018;**82**:714-733. DOI: 10.1016/j.rser.2017.09.081

[3] Liu M, Zhang R, Chen W. Graphene-supported Nanoelectrocatalysts for fuel cells: Synthesis, properties, and applications. Chemical Reviews. 2014;**114**(10):5117-5160. DOI: 10.1021/cr400523y

[4] Su H, Hu YH. Recent advances in graphene-based materials for fuel cell applications. Energy Science & Engineering. 2020;**00**:1-26. DOI: 10.1002/ese3.833

[5] Huang P-H, Kuo J-K, Jiang W-Z, Wu C-B. Simulation analysis of hydrogen recirculation rates of fuel cells and the efficiency of combined heat and power. International Journal of Hydrogen Energy. 2020;**46**(31):16823-16835. DOI: 10.1016/j.ijhydene.2020.08.010

[6] Xia J, Geng Y, Huang S, Chen D, Li N, Xu Q, et al. High-performance anode material based on S and N co-doped graphene/iron carbide nanocomposite for microbial fuel cells. Journal of Power Sources. 2021;**512**:230482. DOI: 10.1016/j.jpowsour.2021.230482

[7] Ghosh A, Verma A. Carbon-polymer composite bipolar plate for HT-PEMFC. Fuel Cells. 2014;**14**(2):259-265. DOI: 10.1002/fuce.201300186

[8] Nemiwal M, Zhang TC, Kumar D. Graphene-based electrocatalysts: Hydrogen evolution reactions and overall water splitting. International Journal of Hydrogen Energy. 2021;**46**(41):21401-21418. DOI: 10.1016/j.ijhydene.2021.04.008

[9] He Y, Wang J, Zhang H, Zhang T, Zhang B, Cao S, et al. Polydopamine-modified graphene oxide nanocomposite membrane for polymer electrolyte membrane fuel cell under anhydrous conditions. Journal of Materials Chemistry A. 2014;**2**(25):9548. DOI: 10.1039/c3ta15301k

[10] Zobir SAM, Rashid SA, Tan T. Chapter 4 - Recent development on the synthesis techniques and properties of graphene derivatives. In: Rashid SA, Othman RNIR, Hussein MZ, editors. Micro and Nano Technologies, Synthesis, Technology and Applications of Carbon Nanomaterials. Elsevier; 2019. pp. 77-107. DOI: 10.1016/B978-0-12-815757-2.00004-8. ISBN 9780128157572

[11] Kakati BK, Ghosh A, Verma A. Efficient composite bipolar plate reinforced with carbon fiber and graphene for polymer electrolyte membrane fuel cell. International Journal of Hydrogen Energy. 2012;**38**(22):9362-9369. DOI: 10.1016/j.ijhydene.2012.11.075

[12] Ren YJ, Anisur MR, Qiu W, He JJ, Al-Saadi S, Singh Raman RK. Degradation of graphene coated copper in simulated polymer electrolyte membrane fuel cell environment: Electrochemical impedance spectroscopy study. Journal of Power Sources. 2017;**362**:366-372. DOI: 10.1016/j.jpowsour.2017.07.041

[13] Jiang X, Drzal LT. Exploring the potential of exfoliated graphene nanoplatelets as the conductive filler in polymeric nanocomposites for bipolar plates. Journal of Power Sources. 2012;**218**:297-306. DOI: 10.1016/j.jpowsour.2012.07.001

[14] Plengudomkit R, Okhawilai M, Rimdusit S. Highly filled graphene-benzoxazine composites as bipolar plates in fuel cell applications. Polymer Composites. 2014;**37**(6):1715-1727. DOI: 10.1002/pc.23344

[15] Iwan A, Malinowski M, Pasciak G. Polymer fuel cell components modified by graphene: Electrodes, electrolytes and bipolar plates. Renewable and Sustainable Energy Reviews. 2015;**49**:954-967. DOI: 10.1016/j.rser.2015.04.093

[16] Ye Y-S, Tseng C-Y, Shen W-C, Wang J-S, Chen K-J, Cheng M-Y, et al. A new graphene-modified protic ionic liquid-based composite membrane for solid polymer electrolytes. Journal of Materials Chemistry. 2011;**21**(28):10448. DOI: 10.1039/c1jm11152c

[17] Qiu X, Dong T, Ueda M, Zhang X, Wang L. Sulfonated reduced graphene oxide as a conductive layer in sulfonated poly(ether ether ketone) nanocomposite membranes. Journal of Membrane Science. 2017;**524**:663-672. DOI: 10.1016/j.memsci.2016.11.064

[18] Holmes SM, Balakrishnan P, Kalangi VS, Zhang X, Lozada-Hidalgo M, Ajayan PM, et al. 2D crystals significantly enhance the performance of a working fuel cell. Advanced Energy Materials. 2016;**7**(5):1601216. DOI: 10.1002/aenm.201601216

[19] Ji Z, Chen J, Guo Z, Zhao Z, Cai R, Rigby MTP, et al. Graphene/carbon structured catalyst layer to enhance the performance and durability of the high-temperature polymer electrolyte membrane fuel cells. Journal of Energy Chemistry. 2022;**75**:399-407. DOI: 10.1016/j.jechem.2022.08.004

[20] Zhang C, Sha J, Fei H, Liu M, Yazdi S, Zhang J, et al. Single-atomic ruthenium catalytic sites on nitrogen-doped graphene for oxygen reduction reaction in acidic medium. ACS Nano. 2017;**11**(7):6930-6941. DOI: 10.1021/acsnano.7b02148

[21] Liu J, Qiao Y, Guo CX, Lim S, Song H, Li CM. Graphene/carbon cloth anode for high-performance mediatorless microbial fuel cells. Bioresource Technology. 2012;**114**:275-280. DOI: 10.1016/j.biortech.2012.02.116

[22] Hur SH, Park J-N. Graphene and its application in fuel cell catalysis: A review. Asia-Pacific Journal of Chemical Engineering. 2012;**8**(2):218-233. DOI: 10.1002/apj.1676

[23] Lee J, Noh S, Pham ND, Shim JH. Top-down synthesis of S-doped graphene nanosheets by electrochemical exfoliation of graphite: Metal-free bifunctional catalysts for oxygen reduction and evolution reactions. Electrochimica Acta. 2012;**313**:1-9. DOI: 10.1016/j.electacta.2019.05.015

[24] Mbayachi VB, Ndayiragije E, Sammani T, Taj S, MbutaER KAU. Graphene synthesis, characterization and its applications: A review. Results in Chemistry. 2021;**3**:100163. DOI: 10.1016/j.rechem.2021.100163

[25] Deepa C, Rajeshkumar L, Ramesh M. Preparation, synthesis, properties and characterization of graphene-based 2D nano-materials for biosensors and bioelectronics. Journal of Materials Research and Technology. 2022;**19**:2657-2694. DOI: 10.1016/j.jmrt.2022.06.023

[26] Liu J, Shuping C, Yanan L, Bijing Z. Progress in preparation, characterization, surface functional modification of graphene oxide: A review. Journal of Saudi Chemical Society. 2022;**26**:101560. DOI: 10.1016/j.jscs.2022.101560

[27] Singh RS, Gautam A, Rai V. Graphene-based bipolar plates for polymer electrolyte membrane fuel cells. Frontiers of Materials Science. 2019;**13**(3):217-241. DOI: 10.1007/s11706-019-0465-0

[28] Tsang ACH, Kwok HYH, Leung DYC. The use of graphene based materials for fuel cell, photovoltaics, and supercapacitor electrode materials. Solid State Sciences. 2017;**67**:A1-A14. DOI: 10.1016/j.solidstatesciences.2017.03.015

[29] Zhu Y, Murali S, Cai W, Li X, Suk JW, Potts JR, et al. Graphene and graphene oxide: Synthesis, properties, and applications. Advanced Materials. 2010;**22**(35):3906-3924. DOI: 10.1002/adma.201001068

[30] Chen D, Feng H, Li J. Graphene oxide: Preparation, functionalization, and electrochemical applications. Chemical Reviews. 2012;**112**(11):6027-6053. DOI: 10.1021/cr300115g

[31] Navalon S, Dhakshinamoorthy A, Alvaro M, Garcia H. Carbocatalysis by graphene-based materials. Chemical Reviews. 2014;**114**(12):6179-6212. DOI: 10.1021/cr4007347

[32] Kong XK, Chen CL, Chen QW. Doped graphene for metal-free catalysis. Chemical Society Reviews. 2014;**43**(8):2841-2857. DOI: 10.1039/c3cs60401b

[33] Ambrosi A, Chua CK, Latiff NM, Loo AH, Wong CHA, Eng AYS, et al. Graphene and its electrochemistry – An update. Chemical Society Reviews. 2016;**45**(9):2458-2493. DOI: 10.1039/c6cs00136j

[34] Ali A, Shen PK. Recent progress in graphene-based nanostructured electrocatalysts for overall water splitting. Electrochemical Energy Reviews. 2020;**3**:370-394. DOI: 10.1007/s41918-020-00066-3

[35] Liang Y, Li Y, Wang H, Zhou J, Wang J, Regier T, et al. Co_3O_4 nanocrystals on graphene as a synergistic catalyst for oxygen reduction reaction. Nature Materials. 2011;**10**(10):780-786. DOI: 10.1038/nmat3087

[36] Fang W, Wang J, Hu Y, Cui X, Zhu R, Zhang Y, et al. Metal-organic framework derived Fe-Co-CN/reduced graphene oxide for efficient HER and OER. Electrochimica Acta. 2021;**365**:137384. DOI: 10.1016/j.electacta.2020.137384

[37] Ghosh A, Basu S, Verma A. Graphene and functionalized graphene supported platinum catalyst for PEMFC. Fuel Cells. 2013;**13**(3):355-363. DOI: 10.1002/fuce.201300039

[38] Perez-Page M, Sahoo M, Holmes SM. Single layer 2D crystals for electrochemical applications of ion exchange membranes and hydrogen evolution catalysts. Advanced Materials Interfaces. 2019;**6**(7):1801838. DOI: 10.1002/admi.201801838

[39] Kim Y, Ketpang K, Jaritphun S, Park JS, Shanmugam SA. Polyoxometalate coupled graphene oxide–Nafion composite membrane for fuel cells operating at low relative humidity. Journal of Materials Chemistry A. 2015;**3**(15):8148-8155. DOI: 10.1039/c5ta00182j

[40] Liu G, Jin W, Xu N. Graphene-based membranes. Chemical Society Reviews.

2015;**44**(15):5016-5030. DOI: 10.1039/c4cs00423j

[41] Liu Y, Min L, Zhang W, Wang Y. High-performance graphene coating on titanium bipolar plates in fuel cells via Cathodic electrophoretic deposition. Coatings. 2021;**11**:437. DOI: 10.3390/coatings11040437

[42] Mahalingam S, Ayyaru S, Ahn YH. Enhanced cathode performance of Fe_2O_3, boron nitride-doped rGO nanosheets for microbial fuel cell applications. Sustainable Energy & Fuels. 2020;**4**:1454-1468. DOI: 10.1039/c9se01243e

[43] Tan L, Pan QR, Wu XT, Li N, Song JH, Liu ZQ. Core@shelled Co/CoO embedded nitrogen-doped carbon Nanosheets coupled graphene as efficient cathode catalysts for enhanced oxygen reduction reaction in microbial fuel cells. ACS Sustainable Chemistry & Engineering. 2019;7:6335-6344. DOI: 10.1021/acssuschemeng.9b00026

[44] Yan Y, Hou Y, Yu Z, Tu L, Qin S, Lan D, et al. B-doped graphene quantum dots implanted into bimetallic organic framework as a highly active and robust cathodic catalyst in the microbial fuel cell. Chemosphere. 2022;**286**:131908. DOI: 10.1016/j.chemosphere.2021.131908

[45] Ren Y, Pan D, Li X, Fu F, Zhao Y, Wang X. Effect of polyaniline-graphene nanosheets modified cathode on the performance of sediment microbial fuel cell. Journal of Chemical Technology & Biotechnology. 2013;**88**(10):1946-1950. DOI: 10.1002/jctb.4146

[46] Kirubaharan CJ, Santhakumar K, Gnana Kumar G, Senthil Kumar N, Jang JH. Nitrogen doped graphene sheets as metal free anode catalysts for the high performance microbial fuel cells. International Journal of Hydrogen Energy. 2015;**40**(38):13061-13070. DOI: 10.1016/j.ijhydene.2015.06.025

[47] Li Z, Yang S, Song Y, Xu H, Wang Z, Wang W, et al. In-situ modified titanium suboxides with polyaniline/graphene as anode to enhance biovoltage production of microbial fuel cell. International Journal of Hydrogen Energy. 2019;**44**(12):6862-6870. DOI: 10.1016/j.ijhydene.2018.12.106

Chapter 7

Biomedical Applications of Chitosan-Coated Nanosheets

Sasireka Rajendran, Vinoth Rathinam, Vasanth Kumar, Manusree Kandasamy, Sharmila Selvi Muthuvel and Shanmugasundari Arumugam

Abstract

Nanotechnology paves the way for the synthesis of novel nanomaterials in one or more dimensions with a size range of less than 100 nm and enhances its specific application because of its unique properties. Nanosheets are a type of layered nanomaterial mostly designed using graphene, poly(L-lactic acid) (PLLA), and molybdenum disulfide by liquid exfoliation method and are of great interest. Nanosheets fabricated could be employed with other materials to enhance their application in diverse areas. Chitosan, a cationic polymer in conjugation with various nanosheets, was designed for various applications like sensors, cancer treatment, drug delivery, and so on. Chitosan-decorated different nanosheets were formulated by various methods, and their diverse application will be focused.

Keywords: graphene oxide, molybdenum disulfide, poly(L-lactic acid), nanosheets, chitosan

1. Introduction

Chitosan, a cationic polymer, recently attracted researchers because of its extensive properties such as nontoxic, biocompatible, biodegradable, and so on. In addition, chitosan possesses excellent film-forming ability, hydrogel, scaffold, nanofibers, and so on. Hence, this is preferred for various applications as an antibacterial agent. Similarly, nanosheets also favor various benefits because of their nano dimension. Chitosan polymer in combination with different nanosheets is fabricated nowadays to improve its applicability. Nanosheets synthesized using different approaches decorated with chitosan are evaluated for various applications such as antifungal effects, sensors, cancer-specific antigen detection, photo-thermal effect in cancer, drug delivery, solvent dehydration, and so on. In this present chapter, chitosan polymer in combination with various nanosheets and its diverse application is discussed.

2. Nanosheets

Nanosheet (NS) is a layered material of various and highly unexploited sources in two-dimension of nanometer sizes in the 1–100 nm thickness range. Atoms and molecules are usually arranged in single or multilayer of two dimensions in the nanosheets [1]. Nowadays, nanotechnology offers a wide choice to researchers to fabricate novel 2D materials over 1D and make them more specific and superior while comparing them with bulk materials. Top-down or bottom-up approaches could be employed for the synthesis of 2D nanosheets [2]. In general, 2D materials possess a high surface area-to-volume ratio and strength, making them more appealing and compatible with specific applications in wide areas such as cell culture, platelet adhesion, drug delivery, and so on [3].

2.1 Different types of nanosheets

Nanosheets of various forms such as graphene oxide (GO), poly(L–lactic acid), and molybdenum disulfide (MoS_2) are available up-to-dately and are currently used in cellular research, drug delivery, multimodal contrast agents, sensors, membranes, optical and electronic devices, nanomedicine, and other specific applications [4–6].

Graphene nanosheets are monolayers of carbon atoms arranged in honeycomb crystal assembly [7] with high electron mobility, thermal stability and conductivity, surface area, and biocompatibility [8] and hence could be used as electrochemical immunosensors for disease diagnosis [9]. Graphene oxide nanosheets help DNA cargo functions and study cellular interactions *in vivo* [10].

Poly(L-lactic acid) (PLLA) nanosheets are known to have anti-adhesive properties and find their application in wound dressing [11]. Better transparency, flexibility, and adhesiveness make it more specific for PLLA nanosheets for burn wounds [12].

Molybdenum disulfide (MoS_2)-based nanosheets are arranged in two-dimensional hexagonal lattices with a single Mo atom. Two S atoms sandwich the Mo atom and result in the formation of a hexagonal honeycomb structural appearance with three layers [6]. Covalent bond formation occurs between the S-Mo-S and makes the nanosheet more stable and compatible with various applications. High surface area, band gap nature, thin size, and better load bearing make it more appealing and enhance its usability.

2.2 Properties of nanosheets

Large contact area, hetero-functionality, noncovalent adhesion, flexibility, and minimum mass introduction are the unique properties of the nanosheet. In addition, high surface area, good mechanical stability, strength, sensitivity, selectivity, and flexibility are of major concern [13].

2.3 Advantages of nanosheets

Nanosheets can be subjected to surface modification very easily, and hence, various alterations could be favored to enhance its property for various applications. Surface structure, charge, hydrophilicity, and hydrophobicity could be altered by surface modification [14].

2.4 Application of nanosheets

Ease of use, flexibility, and modification enhance the specific application of nanosheets as surface-sensitive, substrate material, and in regenerative medicine. In addition, nanosheets are being preferred for other applications such as drug delivery, cell culture, imaging, and sensing [15, 16]. Recent insight into the use of nanosheets with polymer chitosan has received much attention since it has several advantages. Several materials have been combined with chitosan for the fabrication of nanosheets for specific applications [17].

3. Chitosan

Chitin (β-(1–4)-poly-N-acetyl-D-glucosamine) is one of the natural polymer present in the cell walls of exoskeletons of crabs, cuticles of insects, shrimps, and cell walls of fungi. Chitin is deacetylated into chitosan for biomedical applications, using chemicals or with the help of enzymes [18].

Chitosan comprises β-1, 4-linked 2-amino-2-deoxy-β-D-glucose (deacetylated D-glucosamine) and N-acetyl-D-glucosamine units of lower molecular weight (MW) and crystallinity than chitin (MW > 100 kDa) [19]. Moreover, it is structurally similar to cellulose except for the hydroxyl group (−OH) at the C-2 position and is exchanged with the amino group ($-NH_2$) in chitosan. Chitosan is a biocompatible, approved polymer for biomedical applications [20].

Different methods have been followed so far for the preparation of chitosan-based nanocomposites. The solvent casting method, electrospinning, and other methods could be employed for the use of nanocomposites [21]. Chitosan, a cationic deacetylated polymer from chitin, has recently found its application in diverse fields such as drug delivery; antimicrobial activity; excellent film-forming ability, tissue engineering; antitumor, antioxidant, and antifungal activities; wastewater treatment; wound healing; cosmetics; textile and paper; pharmaceutical; and agriculture because of its biocompatibility, nontoxic nature, and structural similarity to natural glycosaminoglycans [22]. Chitosan finds its application in the food industry as an antimicrobial packaging material. In addition, it is added as an emulsifier in the food materials [23].

4. Preparation methods for chitosan-coated nanosheets

4.1 Molybdenum disulfide nanosheets

Sodium molybdate dehydrate (50 mg) and thiourea prepared in deionized water of 50 ml were sonicated at room temperature for 15 minutes, and the pH was maintained at 3.5 using 0.1 M HCl. The mixture was then placed in a Teflon-lined autoclave and subjected to a heating process at 200°C for a period of 18 hours in a hot air oven. After sufficient reaction time, the contents were centrifuged for 10 minutes at 8000 rpm, and the pellet was washed with millipore water and ethanol several times to obtain the black color precipitate and further dried at 80°C under vacuum condition. The final product obtained is molybdenum disulfide nanosheets [24]. **Figure 1** illustrates the proposed methodology of synthesis of molybdenum disulfide nanosheets.

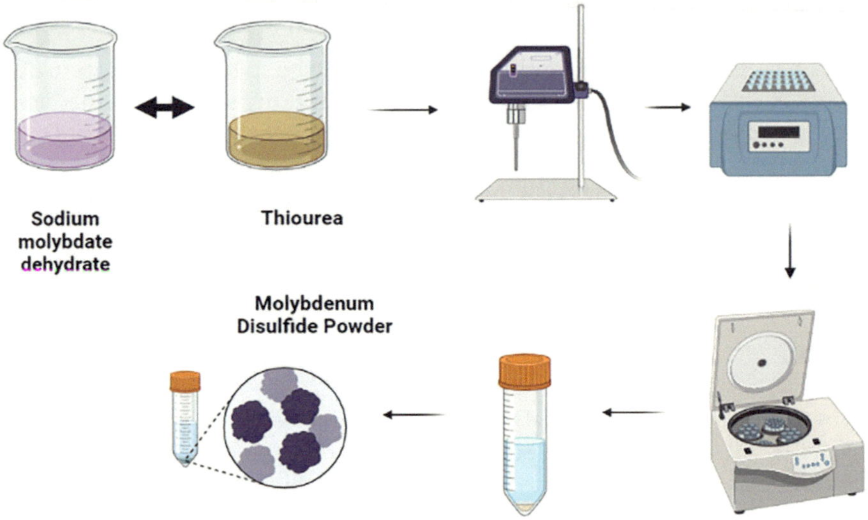

Figure 1.
Proposed strategy of MoS_2 nanosheet synthesis.

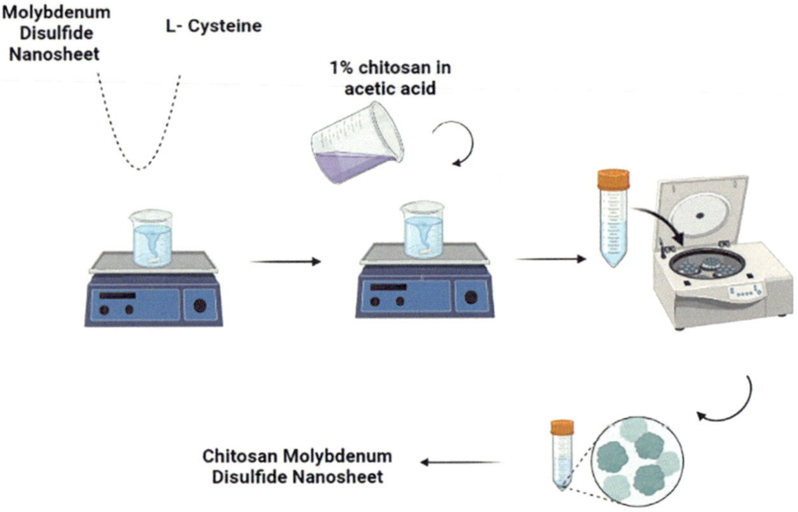

Figure 2.
Strategy involved in the fabrication of chitosan MoS_2 nanosheet.

4.2 Chitosan-based molybdenum disulfide nanosheets

Molybdenum nanosheet (0.3 g) and L–cysteine (30 mg) dissolved in 25 ml of millipore water were stirred continuously for 5 h, and 1% chitosan suspended in acetic acid was added slowly to this mixture at room temperature for the development of chitosan molybdenum disulfide nanosheets. The resultant product obtained after 5 h was collected by centrifuging followed by washing with water and ethanol several times. The recovered product was dried under a vacuum overnight. **Figure 2** represents the strategy involved in the preparation of chitosan

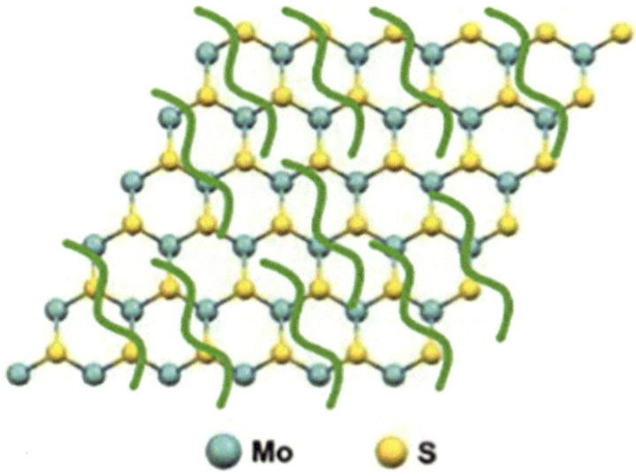

Figure 3.
Chitosan molybdenum disulfide nanosheet.

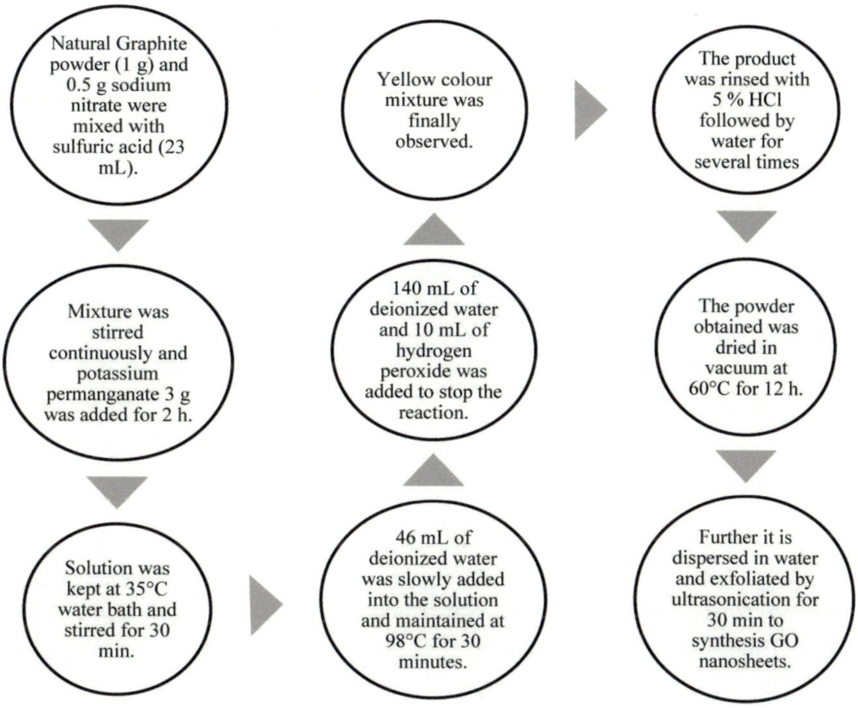

Figure 4.
Steps involved in the synthesis of graphene nanosheet.

molybdenum disulfide nanosheets. **Figure 3** denotes the chitosan molybdenum disulfide nanosheets.

4.3 Graphene nanosheets

Graphene possesses exciting properties such as thermal, optical, electrical, and mechanical properties in its 2D form and is thus preferred for various applications. Graphene in the form of nanosheet decorated with chitosan has multiple benefits. The steps involved in the synthesis of graphene nanosheets are discussed in **Figures 4** and **5** represents graphene nanosheets.

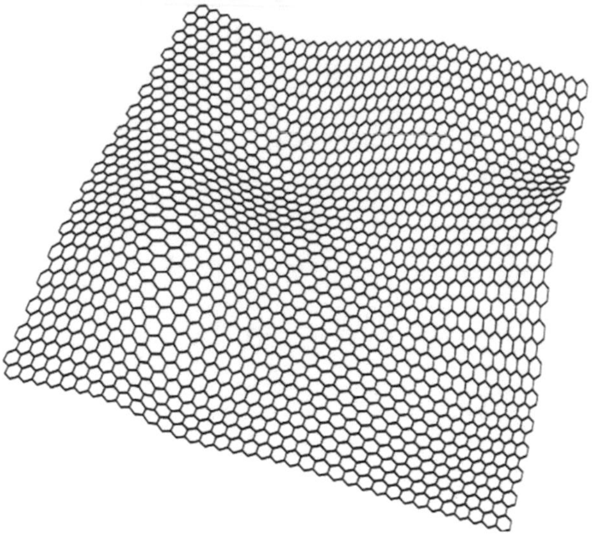

Figure 5.
Graphene nanosheet.

4.4 PLLA nanosheet

PLLA suspended in 1,1,1,3,3,3-hexafluoropropan-2-ol or dichloromethane at 5–50 mg/mL concentration was poured onto silicon dioxide substrate and spin coated for 20 s at 4000 rpm. It was then allowed to heat at 70°C for 90 s. PVA solution was prepared by dissolving in distilled water at 100 mg/mL concentration and dried up at 70°C for 15 min [11]. **Figure 6** represents the fabrication of PLLA nanosheet.

5. Characterization of chitosan-coated nanosheets

Nanosheets fabricated can be characterized by various spectroscopic methods such as FTIR, X-ray Diffraction, and EDX and other microscopic methods such as SEM, FESEM, AFM, TEM, and so on. Nanosheets dimensions could be confirmed while we characterize using different spectroscopic techniques.

6. Applications of chitosan-coated nanosheets

6.1 Antibacterial/antifungal activity

Molybdenum disulfide (MoS_2) nanosheets fabricated using the liquid exfoliation method were laden with chitosan polymer and silver nanoparticles to test the antifungal activity in a study conducted by Zhang et al. [25]. Nanosheets were further characterized by UV-Vis absorption spectroscopy, SEM, and EDS [25]. *In vivo* and *in vitro* studies were conducted to check the efficiency of the synthesized nanosheets. MoS_2-CS-Ag nanosheets were tested against the selected strains such as *Aspergillus niger* and *Saccharomyces uvarum*. Results revealed that better antifungal activity was exhibited with the use of nanosheets both *in vivo* and *in vitro*, and its subsequent application in fruit protection was studied.

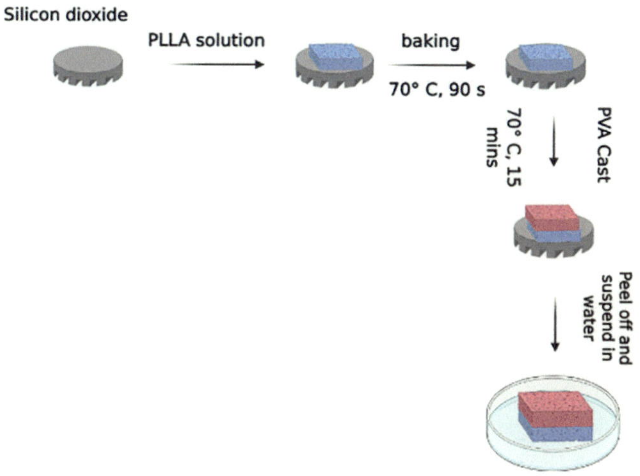

Figure 6.
Proposed mechanism of fabrication of PLLA nanosheet.

Nanosheets of molybdenum disulfide with chitosan and silver nanoparticles exhibit better antifungal activity in a study conducted by Zhang et al. [25]. In this present study, MoS_2 nanosheets were produced by the liquid exfoliation method, and subsequently, chitosan was functionalized with this nanosheet and decorated with Ag NPs. The thickness of the nanosheet was found to be approximately 10 nm. Fungal cultures such as *Aspergillus niger* (ATCC 6275) and *Saccharomyces uvarum* (ATCC 9080) were selected for the antifungal activity assay. Experiments were carried out at different concentrations, and the results revealed that MoS_2 nanosheets loaded with chitosan and Ag-NPs exhibit better antifungal effects when compared with control, and their application in fruit models such as apples and bananas was also examined. Cell morphology, membrane integrity, and cytotoxicity assays are evidence that MoS_2-CS-Ag-NPs nanosheets were found to have promising effects in food preservation.

In another study conducted by Saha et al. [26], chitosan and molybdenum diselenide ($MoSe_2$) nanosheets were synthesized using liquid exfoliation, and their subsequent application in fungal eradication was studied. Antifungal activities of $MoSe_2$/CS nanosheets against variety of unicellular and filamentous fungal strains such as *Cryptococcus gattii* (*C. gattii*, ATCC MYA-4071), *Candida parapsilosis* (*C. parapsilosis*, ATCC 22019), *Issatchenkia orientalis* (*I. orientalis*, ATCC 6258), *Aspergillus fumigatus* (*A. fumigatus* ATCC, MYA-4609), *Candida albicans* (*C. albicans*, ATCC 76485), *Fusarium verticillioides* (*F. verticillioides*, ATCCMYA-3629), *Saccharomyces cerevisiae* (*S. cerevisiae*, ATCC 9763), *Cryptococcus neoformans* (*C. neoformans*, ATCC 208821), *Fusarium falciforme* (*F. falciforme*, ATCC MYA-3636). *C. auris* (0389), *Candida duobushaemulonii* (*C. duobushaemulonii*, 0394), *Candida auris* (*C. auris*, 0386), *Candida haemulonii* (*C. haemulonii*, 0395), *Candida krusei* (*C. krusei*, 0397), *C. auris* (0388), *Krusei ohmeri* (*K. ohmeri*, 0396), *Saccharomyces cerevisiae* (*S. cerevisiae*, 0399), and *Candida lusitaniae* (*C. lusitaniae*, 0398) were studied [26].

The minimum inhibitory concentration (MIC) and minimum fungicidal concentration (MFC) were determined using the $MoSe_2$/CS nanosheets. MIC values for most of the unicellular fungal strains lie in the range of 0.78 and 37.5 μg/ml. Similarly, MFCs range between 0.5 and 75 μg/ml for various filamentous and unicellular strains.

Candida auris strains were also tested against the prepared nanosheets. MICs and MFCs were recorded as 25–50 μm/ml and 37.5 to 150 μm/ml. $MoSe_2$/CS nanosheets were found to have good antifungal activity than graphene-based nanosheets. Cell morphology and the mechanism of killing were also investigated for the selected strains *A. fumigatus* and *C. albicans*. Cell membrane damage, cytoplasmic leakage, depolarization of the membrane, and metabolic inactivation occur as a result of exposure to $MoSe_2$/CS nanosheets. The present study showed promising evidence for the use of $MoSe_2$/CS nanosheets as an antifungal agent.

CS/$MoSe_2$ nanosheets also exhibit better antibacterial activity against selected gram- positive and gram-negative bacteria in a work carried out by Roy et al. [27]. Electrostatic forces and Vanderwaal's interaction between chitosan and molybdenum disulfide nanosheet enhances their application as an antibacterial agent. Strains such as Staphylococcus aureus (S. aureus) and Escherichia coli (E. coli) were tested against the synthesized nanosheets [27] to prove its efficacy. Minimum inhibitory concentration (MIC) and minimum bactericidal concentration (MBC) against the selected strains were also evaluated. Membrane attachment, membrane puncturing, ROS generation, intracellular oxidative stress, membrane depolarization and disruption, and metabolic arrest are the proposed mechanisms that lead to cell death [28].

6.2 Drug delivery

Chitosan functionalized with molybdenum disulfide has promising applications in drug delivery. Chitosan/graphene nanosheets were loaded with doxorubicin for drug delivery in a study conducted by Mousavi et al. [29]. Graphene/N-phthaloylchitosan-graft-poly (methylmethacrylate-block-(poly ethylene glycol methacrylate-random-dimethylamino ethyl methacrylate) GO/CS-g-PMMA-b-(PEGMA-ran-PDMAEMA) nanocomposite was prepared and loaded with the drug to check its efficacy. *In vitro* release, kinetics were also assessed. Promising results revealed that novel nanosheets functionalized with chitosan exhibited better drug delivery efficacy [29].

6.3 Photothermal therapy of cancer

In a study conducted by Rajasekar et al. [30], MoS_2 nanosheet was fabricated at 200°C using the hydrothermal route for 24 h. 1% chitosan solution was coated onto the fabricated nanosheets to form chitosan-coated molybdenum disulfide (CS-MoS_2). Tantalum oxide (TaO_2) was further incorporated into the nanosheets to synthesize TaO_2-CS-MoS_2. Electrostatic interactions between the nanosheets make them suitable for photothermal treatment (PTT). The prepared nanosheets were tested for their biocompatibility, photostability, and cytotoxic effects on the breast cancer cell lines [30].

6.4 Sensors

MoS_2 nanosheets were initially synthesized by ultrasonic exfoliation and subsequently conjugated with chitosan and bismuth film and GC electrode (MoS_2/GC electrode). Further, the prepared nanosheets were modified by functionalization with thiolated poly (aspartic acid) (TPA-MoS_2/GC electrode) in a study conducted by Cao et al. [31]. The TPA-MoS_2/GC sensor was developed to focus on the detection of cadmium ions (Cd^{2+}). TPA-modified MoS_2 nanosheets were found to be an effective and reliable tool to monitor the Cd^{2+} in food and water [31].

6.5 Detection of prostate specific antigen

In a study conducted by Duan et al. [32], MoS_2 quantum dots (MoS_2 QDs) and two-dimensional graphitic carbon nitride (g-C_3N_4) nanosheets were decorated with chitosan and gold nanoparticles and further fabricated with aptamers (MoS_2QDs@g-C_3N_4@CS-AuNPs) to detect prostate-specific antigen (PSA). The construct displayed acts as an electrochemical aptasensor and surface plasmon resonance (SPR) sensor. The limit of detection was found to be 0.71 pg/mL. Enhanced stability, good selectivity, and reproducibility were achieved with the sample tested. Results revealed that it could be used for sensing applications and cancer biomarker prediction [32].

6.6 Dehydration of solvents

Xu et al. [33] fabricated two-dimensional $Ti_3C_2T_x$ MXene nanosheets, which were incorporated with chitosan for efficient solvent dehydration. Organic solvents such as dimethyl carbonate, ethyl acetate, and ethanol were used for evaluating the dehydration performance. Interactions between the chitosan polymer and the hydroxyl groups of MXene improve the infiltration of water molecules across the membrane [33].

6.7 Nitrite oxidation

An inorganic pollutant, nitrite poses a severe hazard to the environment, and it influences the oxidation of hemoglobin and affects the oxygen transfer in the blood. Electrochemical oxidation of nitrite is achieved via the graphene nanosheets and chitosan polymer in a work carried out by Chi et al. [34]. Carbon nanospheres and graphene nanosheet mixture were synthesized and then fabricated with chitosan Prussian blue nanocomposites and glassy carbon electrode (CS@PB/GNS-CNS/GCE) for the nitrite oxidation. Electrostatic interactions between the chitosan/Prussian blue nanocomposite and graphene nanosheets/carbon nanospheres make it an excellent redox mediator for nitrite detection. It is evident from the results that sensors could be developed based on their detection limit and could be used in food samples [34].

6.8 Implantation in heart

In a study conducted by Saravanan et al. [35], graphene oxide nanosheets loaded with gold nanoparticles were incorporated into a chitosan scaffold to check its efficacy in infarcted hearts. Ventricular contractility and its function subsequently improved after implantation. The synthesized scaffold exhibited a better swelling nature and controlled degradation. The attachment of the cells to the scaffold and its growth is enhanced upon the use of nanosheets. The novel scaffold and its application in heart implantation for ventricular function are assessed. Results revealed that the expression of connexin 43 also significantly improved. Immunohistochemistry and immunocytochemistry studies showed prominent results for the implantation of the scaffolds [35].

6.9 Corrosion resistance

Huang et al. [36] developed a chitosan (CS)/boron nitride nanosheet (BNNS) composite, which was fabricated on the surface of Mg-Zn-Y-Nd-Zr alloy by electrodeposition strategy. The construct was designed to diminish the magnesium alloy degradation so that it could be used in bone implants since it possesses antibacterial activity. BNNS has significant changes while coated with chitosan and affords better shelf life to substrate. *E. coli* and *S. aureus* were tested to prove the fabricated nanosheet has antibacterial activity. Cytocompatibility tests revealed that the CS-BNNS coating has no adverse effect on the adhesion and proliferation of mouse osteoblasts cells. Thus, the fabricated CS-BNNS composite coating could be a promising application in surface modification for Mg-based bone implants [36].

6.10 Membranes as proton exchangers

In a study carried out by Divya et al. [37], MoS_2 nanosheets were fabricated initially by exfoliation method, and chitosan in different ranges were used for the development of the chitosan-MoS_2 nanocomposites acting as proton exchange membranes because of their unique nature such as proton conductivity, water absorbing capacity, and ion exchange capacity. Characterization of the chitosan-MoS_2 nanosheet membranes was performed using FT-IR, XRD, FESEM-EDX, and AFM. Contact angle, uptake of water, swelling ratio, ion exchange, and proton conductivity were analyzed. Among the various composition selected, 0.75% of chitosan membranes exhibited high membrane selectivity and proton conductivity [37].

6.11 Detection of mutated DNA

Graphene nanosheets modified with chitosan were placed on the surface of the carbon electrode (CMG electrode) for the electrochemical analysis of mutations in DNA. Nanosheets were initially prepared by Hummers methods and subsequently characterized by FT-IR, Raman spectroscopy, and transmission electron microscopy by Alwarappan et al. [38]. The main advantage of this nanosheet is that attachment of single-stranded polynucleotide in the CMG electrode helps in the detection of complementary strands and finds the mutation [38].

Carboxymethyl cellulose-chitosan-montmorillonite nanosheets were synthesized for the remediation of dye effluent by Wang et al. [39]. TiO_2@MMTNS/CMC/CS nanocomposite was studied for its absorption of methylene blue dye [39]. Cyanometallate/chitosan nanosheet acts as a catalyst to enhance the polysulfide redox reaction in batteries in a work conducted by Fang et al. [40]. Chitosan nanosheets/Honey compounds were tested for wound healing effects in male BALB/c Mice. The wound injury model was developed in adult mice. Animal models were divided into five different groups for the experimental study. Control group, polyethylene glycol (PEG), CNs treated, honey, and CNs dissolved in PEG or honey. While compared with the control group, CNs-treated mice model showed significant results. Better wound healing is achieved upon treatment especially when chitosan nanosheets treated with honey [41].

7. Conclusion

In today's world field of nanotechnology, especially nanosheets, has a promising impact in cellular research, biomedicine, regenerative medicine, and tissue engineering. Fabrication of nanosheets and their application in in-vivo conditions is still challenging. Novel nanosheets with advanced features can be fabricated to address the various applications. Despite the challenges faced, nanosheets could be employed with various modifications and can be used for the modern era. Chitosan-coated nanosheets could be the alternate choice for targeting multiple applications because of their attractive features.

Acknowledgements

The authors would like to acknowledge the institution for their support.

Conflict of interest

The authors declare no conflict of interest.

Author details

Sasireka Rajendran[1*], Vinoth Rathinam[2], Vasanth Kumar[1], Manusree Kandasamy[1], Sharmila Selvi Muthuvel[1] and Shanmugasundari Arumugam[1]

1 Mepco Schlenk Engineering College, Sivakasi, India

2 P.S.R. Engineering College, Sivakasi, India

*Address all correspondence to: sasirekabt@mepcoeng.ac.in

References

[1] Wang G, Yang J, Park J, Gou X, Wang B, Liu H, et al. Facile synthesis and characterization of graphene nanosheets. The Journal of Physical Chemistry C. 2008;**112**(22):8192-8195

[2] Zhang X, Lai Z, Ma Q, Zhang H. Novel structured transition metal dichalcogenide nanosheets. Chemical Society Reviews. 2018;**47**(9):3301-3338

[3] Chhowalla M, Liu Z, Zhang H. Two-dimensional transition metal dichalcogenide (TMD) nanosheets. Chemical Society Reviews. 2015;**44**(9):2584-2586

[4] Zhang S, Sunami Y, Hashimoto H. Mini review: Nanosheet technology towards biomedical application. Nanomaterials. 2017;**7**(9):246

[5] Fujie T, Okamura Y, Takeoka S. Ubiquitous transference of a free-standing polysaccharide nanosheet with the development of a nano-adhesive plaster. Advanced Materials. 2007;**19**(21):3549-3553

[6] Fujie T, Mori Y, Ito S, Nishizawa M, Bae H, Nagai N, et al. Micropatterned polymeric nanosheets for local delivery of an engineered epithelial monolayer. Advanced Materials. 2014;**26**(11):1699-1705

[7] Chaiyakun S, Witit-Anun N, Nuntawong N, Chindaudom P, Oaew S, Kedkeaw C, et al. Preparation and characterization of graphene oxide nanosheets. Procedia Engineering. 2012;**32**:759-764

[8] Chhowalla M, Shin HS, Eda G, Li L-J, Loh KP, Zhang H. The chemistry of two-dimensional layered transition metal dichalcogenide nanosheets. Nature Chemistry. 2013;**5**(4):263-275

[9] Wang L-C, Bao S-K, Luo J, Wang Y-H, Nie Y-C, Zou J-P. Efficient exfoliation of bulk MoS2 to nanosheets by mixed-solvent refluxing method. International Journal of Hydrogen Energy. 2016;**41**(25):10737-10743

[10] Guo H-L, Wang X-F, Qian Q-Y, Wang F-B, Xia X-H. A green approach to the synthesis of graphene nanosheets. ACS Nano. 2009;**3**(9):2653-2659

[11] Miyazaki H, Kinoshita M, Saito A, Fujie T, Kabata K, Hara E, et al. An ultrathin poly (l-lactic acid) nanosheet as a burn wound dressing for protection against bacterial infection. Wound Repair and Regeneration. 2012;**20**(4):573-579

[12] Niwa D, Koide M, Fujie T, Goda N, Takeoka S. Application of nanosheets as an anti-adhesion barrier in partial hepatectomy. Journal of Biomedical Materials Research Part B: Applied Biomaterials. 2013;**101**(7):1251-1258

[13] Rodríguez-San-Miguel D, Montoro C, Zamora F. Covalent organic framework nanosheets: Preparation, properties and applications. Chemical Society Reviews. 2020;**49**(8):2291-2302

[14] Jin X, Gu T-H, Lee K-G, Kim MJ, Islam MS, Hwang S-J. Unique advantages of 2D inorganic nanosheets in exploring high-performance electrocatalysts: Synthesis, application, and perspective. Coordination Chemistry Reviews. 2020;**415**:213280

[15] Sharafeldin M, Bishop GW, Bhakta S, El-Sawy A, Suib SL, Rusling JF. Fe3O4 nanoparticles on graphene oxide sheets for isolation and ultrasensitive amperometric detection of cancer biomarker proteins. Biosensors and Bioelectronics. 2017;**91**:359-366

[16] Siasar KM. Diagnosis GLY120 antigen for the blood and breast cancers using graphene Nanosheet. International Journal of Nanoscience and Nanotechnology. 2017;**13**(4):327-333

[17] Chen Y, Liang Y, Wang L, Guan M, Zhu Y, Yue X, et al. Preparation and applications of freestanding Janus nanosheets. Nanoscale. 2021;**13**(36):15151-15176

[18] Roberts GA, Roberts GA. Chitin Chemistry. New York City: Springer; 1992

[19] Shamshina JL, Berton P, Rogers RD. Advances in functional chitin materials: A review. ACS Sustainable Chemistry & Engineering. 2019;**7**(7):6444-6457

[20] Hirano S. Chitin biotechnology applications. Biotechnology Annual Review. 1996;**2**:237-258

[21] Foster A, Webber J. Chitin. In: Advances in Carbohydrate Chemistry. Vol. 15. Amsterdam, Netherlands: Elsevier; 1961. pp. 371-393

[22] Kumar MNR. A review of chitin and chitosan applications. Reactive and Functional Polymers. 2000;**46**(1):1-27

[23] Rinaudo M. Chitin and chitosan: Properties and applications. Progress in Polymer Science. 2006;**31**(7):603-632

[24] Saada I, Bissessur R. Nanocomposite materials based on chitosan and molybdenum disulfide. Journal of Materials Science. 2012;**47**:5861-5866

[25] Zhang W, Mou Z, Wang Y, Chen Y, Yang E, Guo F, et al. Molybdenum disulfide nanosheets loaded with chitosan and silver nanoparticles effective antifungal activities: In vitro and in vivo. Materials Science and Engineering: C. 2019;**97**:486-497

[26] Saha S, Gilliam MS, Wang QH, Green AA. Eradication of fungi using MoSe2/chitosan nanosheets. ACS Applied Nano Materials. 2022;**5**(1):133-148

[27] Roy S, Deo K, Singh KA, Lee HP, Jaiswal A, Gaharwar AK. Nano-bio interactions of 2D molybdenum disulfide. Advanced Drug Delivery Reviews. 2022;**187**:114361

[28] Cao W, Yue L, Wang Z. High antibacterial activity of chitosan–molybdenum disulfide nanocomposite. Carbohydrate Polymers. 2019;**215**:226-234

[29] Mousavi SM, Babazadeh M, Nemati M, Es' haghi M. Doxorubicin-loaded biodegradable chitosan–graphene nanosheets for drug delivery applications. Polymer Bulletin. 2022;**79**(8):6565-6580

[30] Rajasekar S, Martin EM, Kuppusamy S, Vetrivel C. Chitosan coated molybdenum sulphide nanosheet incorporated with tantalum oxide nanomaterials for improving cancer photothermal therapy. Arabian Journal of Chemistry. 2020;**13**(3):4741-4750

[31] Cao Q, Xiao Y, Huang R, Liu N, Chi H, Lin C-T, et al. Thiolated poly (aspartic acid)-functionalized two-dimensional MoS 2, chitosan and bismuth film as a sensor platform for cadmium ion detection. RSC Advances. 2020;**10**(62):37989-37994

[32] Duan F, Zhang S, Yang L, Zhang Z, He L, Wang M. Bifunctional aptasensor based on novel two-dimensional nanocomposite of MoS2 quantum dots and g-C3N4 nanosheets decorated with chitosan-stabilized Au nanoparticles for selectively detecting prostate specific antigen. Analytica Chimica Acta. 2018;**1036**:121-132

[33] Xu Z, Liu G, Ye H, Jin W, Cui Z. Two-dimensional MXene incorporated chitosan mixed-matrix membranes for efficient solvent dehydration. Journal of Membrane Science. 2018;**563**:625-632

[34] Cui L, Zhu J, Meng X, Yin H, Pan X, Ai S. Controlled chitosan coated Prussian blue nanoparticles with the mixture of graphene nanosheets and carbon nanoshperes as a redox mediator for the electrochemical oxidation of nitrite. Sensors and Actuators B: Chemical. 2012;**161**(1):641-647

[35] Saravanan S, Sareen N, Abu-El-Rub E, Ashour H, Sequiera GL, Ammar HI, et al. Graphene oxide-gold nanosheets containing chitosan scaffold improves ventricular contractility and function after implantation into infarcted heart. Scientific Reports. 2018;**8**(1):15069

[36] Huang W, Mei D, Qin H, Li J, Wang L, Ma X, et al. Electrophoretic deposited boron nitride nanosheets-containing chitosan-based coating on Mg alloy for better corrosion resistance, biocompatibility and antibacterial properties. Colloids and Surfaces A: Physicochemical and Engineering Aspects. 2022;**638**:128303

[37] Divya K, Rana D, Alwarappan S, Saraswathi MSSA, Nagendran A. Investigating the usefulness of chitosan based proton exchange membranes tailored with exfoliated molybdenum disulfide nanosheets for clean energy applications. Carbohydrate Polymers. 2019;**208**:504-512

[38] Alwarappan S, Cissell K, Dixit S, Li C-Z, Mohapatra S. Chitosan-modified graphene electrodes for DNA mutation analysis. Journal of Electroanalytical Chemistry. 2012;**686**:69-72

[39] Wang W, Ni J, Chen L, Ai Z, Zhao Y, Song S. Synthesis of carboxymethyl cellulose-chitosan-montmorillonite nanosheets composite hydrogel for dye effluent remediation. International Journal of Biological Macromolecules. 2020;**165**:1-10

[40] Fang Z, Hu X, Shu C, Jian J, Liu J, Yu D. Crosslinked cyanometallate–chitosan nanosheet assembled aerogels as efficient catalysts to boost polysulfide redox kinetics in lithium–sulfur batteries. Journal of Materials Chemistry A. 2020;**8**(37):19262-19268

[41] Askari M, Afshar M, Khorashadizadeh M, Zardast M, Naghizadeh A. Wound Healing Effects of Chitosan Nanosheets/Honey Compounds in Male BALB/c Mice. The International Journal of Lower Extremity Wounds. 2022